BIOCHEMISTRY RESEARCH TRENDS

FUMONISINS

NATURAL OCCURRENCE, MANAGEMENT PRACTICES AND HEALTH CONCERNS

BIOCHEMISTRY RESEARCH TRENDS

Additional books in this series can be found on Nova's website
under the Series tab.

Additional e-books in this series can be found on Nova's website
under the e-book tab.

BIOCHEMISTRY RESEARCH TRENDS

FUMONISINS

NATURAL OCCURRENCE, MANAGEMENT PRACTICES AND HEALTH CONCERNS

CRAIG M. EVANS
EDITOR

New York

NOTICE TO THE READER

Library of Congress Cataloging-in-Publication Data

ISBN: 978-1-63482-789-8

Published by Nova Science Publishers, Inc. † New York

CONTENTS

Preface — vii

Chapter 1 — Fumonisins: Occurrence, Prevention and Health Impacts — 1
Milena Veronezi Silva,
Simone Aparecida Galerani Mossini
and Miguel Machinski Junior

Chapter 2 — Fumonisin B_1: A Potential Risk Factor for Neural Tube Defects — 27
Ana-Marija Domijan, Maja Šegvić Klarić,
Tihana Žanić Grubišić and Lada Rumora

Chapter 3 — State-of-the-Art in the Analysis of Fumonisins by Liquid Chromatography-Mass Spectrometry — 41
Paolo Lucci, Rocío Castro-Ríos
and Oscar Núñez

Index — 99

PREFACE

The genus Fusarium are among the most common contaminants of cereals and their products, and many species have acquired additional importance because they have been shown to produce mycotoxins (fumonisins), causing both animal and human diseases. Fusarium spp. molds and fumonisins were also found to be contaminants on various commodities around the world. The authors discuss the many natural forms of fumonisins and exposure to these mycotoxins, which can cause various adverse health effects in animals and humans. The state-of-the-art liquid chromotography-mass spectrometry techniques for the analysis and characterization of fumonisins in food-based products and other matrices are reviewed as well.

Chapter 1 – Fumonisins are mycotoxins generally produced by *Fusarium verticillioides* and *Fusarium proliferatum*. Maize is the most important source of fumonisin contamination in foods. There are many naturally forms of fumonisins and fumonisin B_1 is the most abundant naturally occurring analogue in maize followed by fumonisin B_2 and fumonisin B_3. Dietary exposure to these mycotoxins can cause various adverse health effects in animals and humans, including equine leukoencephalomalacia, porcine pulmonary edema and neural tube defects in experimental animals. Long-term exposure induces liver and kidney tumors in rats and epidemiological studies have linked consumption of fumonisin contaminated maize with esophageal cancer in human populations. The natural occurrence of mycotoxins in cereals, in addition to health problems, can cause great economic losses throughout the world. Several countries have implemented laws that limit the maximum levels of fumonisin and other mycotoxins that are allowed in food for human and animal consumption, but the good agricultural practices are the main line of defense against contamination of cereals by mycotoxins.

Chapter 2 – Fumonisin B_1 (FB_1) is a mycotoxin and a secondary product of *Fusarium* spp. molds. The predominant producer of FB_1 is *F. verticillioides* (formerly known as *F. moniliforme*). *Fusarium* spp. molds and FB_1 are common contaminants of maize worldwide. The toxicity of FB_1 was first observed in farm animals; in horses, exposure to FB_1-contaminated feed has been associated with equine leukoencephalomalacia, and in pigs with pulmonary edema. For experimental animals, FB_1 is hepatotoxic, nephrotoxic and carcinogenic. In some regions of the world where maize is a staple food, increased incidences of human esophageal and liver cancers as well as increased frequency of neural tube defects (NTDs) have been observed and related to FB_1 exposure. NTDs are congenital malformations that occur as a result of failure of neural tube closure during early embryogenesis; in humans, the most common NTDs are anencephaly and spina bifida. Several lines of evidence point to an association between FB_1 and development of NTDs. It has been well-established that FB_1, due to its structural similarities with sphingoid bases sphinganine and sphingosine, alters sphingolipid metabolism. That in turn affects cell signal transduction pathways and folate receptor functioning. Therefore, FB_1-provoked impairment of sphingolipid metabolism has been suggested as a possible mechanism by which FB_1 induces the development of NTDs. Experimental studies have confirmed the embryotoxicty and teratogenicity of FB_1. Epidemiological studies demonstrated that maternal exposure to FB_1 (through consumption of FB_1-contaminated maize-based food) could be implicated in higher NTD occurrences. In this review, data associating exposure to FB_1 with the development of NTDs are presented.

Chapter 3 – Fumonisins are mycotoxins produced by various *Fusarium* spp. Fungi (mainly *Fusarium verticillioides* and *Fusarium proliferatum*) commonly found in maize. They were first discovered in the late 80s by Gelderblon et al., who studied the incidence rate of esophageal cancer in the South African population. More than ten different fumonisins have been isolated and characterized. Of these, fumonisins B_1, B_2 and B_3 (FB_1, FB_2 and FB_3, respectively) are the major ones naturally occurring in maize or maize-based products. Fumonisins are toxic toward many different animals. Known animal diseases caused by these mycotoxins are hepatocarcinogenic effects of rats and mice, equine leukoencephalomalacia in horses, and pulmonary edema in pigs. However, their toxicity toward humans is still unclear. There is some evidence that fumonisins might be involved in the formation of esophageal cancer and might have a negative effect on neural tube development in embryos. The International Agency for Research on Cancer classified FB_1 as a Group 2B carcinogen (possibly carcinogenic to humans). For these reasons, maximum levels for fumonisin contamination have been decreed by many countries. For instance, in the European Union, the maximum permitted levels for fumonisins (sum of FB_1 and FB_2) in maize and its derived products are 0.2 to 4 mg/Kg. Therefore, the development of analytical methodologies for the routine control of fumonisins in food is necessary. Various methodologies have been employed for the analysis of fumonisins. Although fumonisins molecules neither selectively absorb UV light nor exhibit any fluorescence, since they lack chromophores in their molecular structure, liquid chromatography (LC) with fluorescence detection have been employed after fumonisin derivatization into compounds that exhibit fluorescence. But nowadays, liquid chromatography-mass spectrometry (LC-MS(/MS)) and liquid chromatography-high resolution mass spectrometry (LC-HRMS) are the techniques of choice for the analysis and characterization of fumonisins in foodstuffs. However, it should be pointed out that regardless of the employed detection system, sample preparation procedures (especially extraction and clean-up procedures) is maybe the most important issue in the determination of fumonisins. Various extraction techniques have been reported for the clean-up and extraction of fumonisins including simple solid-liquid extraction (SLE), solid-phase extraction (SPE), immunoaffinity procedures, the use of molecularly imprinted polymers (MIPs), and even QuEChERS methods. This chapter will review the state-of-the-art of liquid chromatography-mass spectrometry techniques for the analysis and characterization of fumonisins in food-based products and other matrices. Because of their importance in fumonisin analysis, commonly used as well as novel sample treatment procedures will be addressed. LC-MS chromatographic conditions, ionization sources, and MS and HRMS analyzers frequently used, as well as strategies for the structural characterization and the qualitative and quantitative analysis of fumonisins will be discussed by means of relevant applications. Coverage of all kind of applications is beyond the scope of the present contribution, so we will focus on the most relevant applications published in the last years.

In: Fumonisins
Editor: Craig M. Evans

ISBN: 978-1-63482-789-8
© 2015 Nova Science Publishers, Inc.

Chapter 1

FUMONISINS: OCCURRENCE, PREVENTION AND HEALTH IMPACTS

Milena Veronezi Silva,
Simone Aparecida Galerani Mossini
and Miguel Machinski Junior
Laboratory of Toxicology, Department of Basic Health Sciences,
State University of Maringa, Maringa, Brazil

ABSTRACT

Fumonisins are mycotoxins generally produced by *Fusarium verticillioides* and *Fusarium proliferatum*. Maize is the most important source of fumonisin contamination in foods. There are many naturally forms of fumonisins and fumonisin B_1 is the most abundant naturally occurring analogue in maize followed by fumonisin B_2 and fumonisin B_3. Dietary exposure to these mycotoxins can cause various adverse health effects in animals and humans, including equine leukoencephalomalacia, porcine pulmonary edema and neural tube defects in experimental animals. Long-term exposure induces liver and kidney tumors in rats and epidemiological studies have linked consumption of fumonisin contaminated maize with esophageal cancer in human populations. The natural occurrence of mycotoxins in cereals, in addition to health problems, can cause great economic losses throughout the world. Several countries have implemented laws that limit the maximum levels of fumonisin and other mycotoxins that are allowed in food for human and animal consumption, but the good agricultural practices are the main line of defense against contamination of cereals by mycotoxins.

Keywords: Fumonisins, *Fusarium*, maize, mycotoxins, occurrence

INTRODUCTION

The genus *Fusarium* are among the most common contaminants of cereals and their products, and many species have acquired additional importance because they have been shown to produce mycotoxins, causing both animal and human diseases [1, 2].

Fusarium species are widely distributed in soil and on organic substrates, especially in the tropics and humid temperate areas of the world. On the last decades, over 1000species of *Fusarium* were described, which led to many of these species were synonyms, in view of the great variability of *Fusarium*in different environments and substrates [3, 4].

Fusarium contains important mycotoxin-producing species. Fumonisins are mycotoxins mainly produced by *Fusarium* species, especially *Fusarium verticillioides* (Sacc.) Nirenberg and *Fusarium proliferatum* (Matsushima) Nirenberg [5–7].

There are many naturally forms of fumonisins, since they have been isolated in 1988, 28 analogs have been described [8]. Fumonisins have been isolated from cultures and maize-based food and they form chemically related groups: A, B, C and P (Figure 1) [9, 10].

Fumonisins are polar compounds, which hinders their study, in contrast to the other mycotoxins, these toxins are soluble in water and aqueous solutions of methanol and acetonitrile, but they are not soluble in non-polar solvents [11, 12].

The pure substance of fumonisin B_1 (FB_1) is a white hygroscopic powder, its empirical formula $C_{34}H_{59}NO_{15}$ is the diester of propane-1, 2, 3-tricarboxylic acid and 2-amino-12,16-dimethyl-3,5,10,14,15-pentahydroxyeicosane. It has a molecular weight of 721 and contains 1024 different stereoisomers [12, 13].

Fumonisin B_2 (FB_2) is the C-10 deoxy analog of fumonisin B_1, in which the C-14 and C-15 hydroxyl groups are esterified by the terminal carboxy group of propane-1, 2, 3-tricarboxylic acid [12].

Fumonisin B_1 is the most abundant of the fumonisin family and usually accounts for 70-80% of the total fumonisin content in naturally contaminated foods, followed by fumonisin B_2 and fumonisin B_3, they generally make up 15-25 and 3-8%, respectively [9, 12].

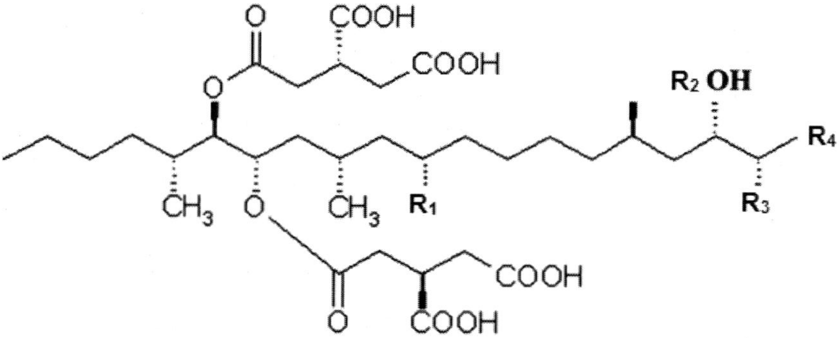

Fumonisins	R_1	R_2	R_3	R_4
Fumonisin A_1	OH	OH	$NHCOCH_3$	CH_3
Fumonisin A_2	H	OH	$NHCOCH_3$	CH_3
Fumonisin A_3	OH	H	$NHCOCH_3$	CH_3
Fumonisin B_1	OH	OH	NH_2	CH_3
Fumonisin B_2	H	OH	NH_2	CH_3
Fumonisin B_3	OH	H	NH_2	CH_3
Fumonisin C_1	OH	OH	NH_2	H
Fumonisin P_1	OH	OH	3HP	CH_3
Fumonisin P_2	H	OH	3HP	CH_3
Fumonisin P_3	OH	H	3HP	CH_3

Figure 1. The chemical structures of the fumonisins.

From a toxicological perspective, fumonisin B_1 is the most important fumonisin and occur in higher levels than others, naturally contaminated maize grains can contain FB_1 and FB_2 at a ratio 3:1 and FB_1/FB_3 at a ratio of 12:1 [12, 14].

OCCURRENCE OF FUMONISINS IN FOOD

Fusarium toxins, after aflatoxins, are the most-often reported mycotoxins in raw agricultural commodities. Their occurrence, frequency and level of the contamination have gained global attention over the last decade.

Table 1. Recent occurrence of Fumonisins in foods

Samples/Number of samples	Origin	Year of collection	Positive samples (%)	Levels range of positive samples (μg/Kg or μg/L)	Reference
Maize (146)	China	2012	39.7%	FB_1 <13-15,221 FB_2 <10-7141	[17]
Cereals and cereal-based foods (290)	Slovenia	2008-2012	2.8%	FB_1+FB_2 <200-27,483	[18]
Barley and malt (52)	Czech Republic	2012	5.8%	FB_1 <20 FB_2 <20-15.64	[19]
Maize products (50)	Brazil	-	60%	FB_1 537.4 (mean)	[20]
Sorghum (10)	Belgium/Germany	-	40%	FB_1 <50-97 FB_2 <50 FB_3 < 50	[21]
Extruded dry dog food (48)	Italy	-	88%	FB_1 <5-1503 FB_2 <5-388	[22]
Soy (36)	Italy	2008-2010	30.6%	FUM 100-2500	[23]
Maize kernels (60)	Tanzania	-	73%	FB_1 16-18184 FB_2 178–38217	[24]
Natural bottled mineral waters and spring waters (26)	Portugal	2012-2013	-	n.d.	[25]
Craft beer (53)	Brazil	2014	15.1%	FB_1 29-285	[26]
Common wheat (135) Durum wheat (40)	Argentina	2011	93%	FB_1 0.15-1304.39 FB_2 0.25-46.94	[27]
Maize (90)	Serbia	2012	100%	FB_1+FB_2+FB_3 520.2–5800	[28]
Breast milk (131)	Northern Tanzania	-	44.3%	FB_1 6.57-471.05	[29]
Traditional beer (9)	Malawi	2012	100%	FB_1 1522 (mean) FB_2 251 (mean) FB_3 229 (mean)	[30]
Rice (20)	Brazil	2011	5%	FB_1 258.7*	[31]
Dried fruits (228)	Tunisia/ Spain	2012-2013	-	n.d.	[32]
Wheat (86)	Syria/Italy	2009-2010	4.7%	FB_1 <LOQ-6 FB_2 <LOQ-12	[33]
Pseudocereais, spelt and rice (45)	Spain	2012	-	n.d.	[34]

Samples/Number of samples	Origin	Year of collection	Positive samples (%)	Levels range of positive samples (µg/Kg or µg/L)	Reference
Wheat (80)	Morocco	-	-	n.d.	[35]
Maize-based foods (120)	Brazil	2011-2012	72.5%	FB_1 33.0-1208.6	[36]
Gluten-free products and maize flour (376)	Italy	-	29%	FB_1 +FB_2 42-761	[37]
Sorghum and finger millet (104)	Ethiopia	-	14.3% (sorghum) 45.5% (finger millet)	FB_1 <LOQ-49.2 FB_2 <LOQ-16.1 FB_3 <LOQ-6.29	[38]
Red wine (280)	Netherlands France, Italy, Spain, Argentina, Australia, Chile, New Zealand, South Africa and United States	2008-2012	-	n.d.	[39]
Maize flour (?)	Tanzania	-	-	FB_1 57-1672	[40]
Chilli (121)	Sri Lanka and Belgium	2012-2013	15%	FB_2 <64.2-87.1	[41]
Black pepper (82) and white pepper (11)	Sri Lanka	-	10% of black pepper	FB_1 <LOQ-134.5	[42]
Nonfermented and fermented feedingstuffs, feedingstuffs supplements, and complex compound feeds (343)	Czech Republic and United Kingdom	2008-2012	-	FB_1 <LOQ-3719 FB_2 <LOQ-542 FB_3 <LOQ-241	[43]
Maize, groundnut, soybean and food-based (125)	Cameroon	2011	-	FB_1 0.4-2313 FB_2 <LOQ-572 FB_3 <LOQ-157 FB_6 <LOQ-4368	[44]
24 different food matrices (240)	Spain	-	6.7%	FB_1 <3.5-13	[45]
Organic and non-organic products (1250)	France, Germany and Spain	2009-2012	16.2%	FB_1 20.2-1201.7 FB_2 21.7-1010.5	[46]
Milk thistle (*Silybummarianum*) (7)	Spain	-	-	n.d.	[47]
Maize germ, refined corn oil and corn oil margarine (74)	Spain	2011-2012	100% maize germ samples, 4% in corn oil and 8% in margarine	FB_1 <100-1553 FB_2 <100-1403	[48]
Rice (9)	Brazil	2011	77.8%	FB_1 30-170	[49]

Table 1. (Continued)

Samples/Number of samples	Origin	Year of collection	Positive samples (%)	Levels range of positive samples (μg/Kg or μg/L)	Reference
Pu-erh tea (36)	China	-	-	n.d.	[50]
Ginseng (43)	China	-	-	n.d.	[51]
Maize, wheat, barley and oat (181)	Croatia	2011	90% Maize, 39%wheat, 15% barley and 6% oat	FUM 25-4438	[52]
Snacks (21)	Nigeria	-	4.8%	FB_1<70 FB_2<100 FB_3<90	[53]
Herbs and spices (79)	Poland	-	39.2%	FUM 5.29–18.96	[54]
Yam flour (100)	Nigeria	2009	32% of white yam 5% of water yam	FB_1 <0.5-91 FB_2 <1-32	[55]
Maize (143)	Italy	-	-	FB_1 +FB_1 417-11845	[56]
Maize (3246)	Argentina	1999-2010	90-100%	FB_1 <18-498,212	[57]
Grain, feed and other feed commodities (1468)	Asian-Oceania region	2010	46.3%	FB_1 <100-19278 FB_2 <100-6684	[58]
Ethnic food (35) gluten-free food (18)	Spain	2009	51% ethnic food and 33.3% gluten-free food	FB_1 5.70-547 FB_2 3.57-135	[59]
Wheat flour (15)	Serbia	2011	-	n.d.	[60]
Maize, wheat and oat (15)	Italy	2010-2011	26.7%	FB_1+FB_2 47–8150	[61]
Maize-based products (100)	Brazil	2007-2010	82%	FB_1 81-3462 FB_2 45-886	[62]
Raw and processed cereals (58)	Tunisia	2009	5%	FB_1 6.4-120	[63]
Maize ears(40)	Brazil	2009	100%	FUM230-6450	[64]
Baby foods (35)	Spain	2010	11.4%	FB_1 75-100 FB_2 75*	[65]
Tiger-nuts (83)	Spain	2010-2011	-	n.d.	[66]

Samples/Number of samples	Origin	Year of collection	Positive samples (%)	Levels range of positive samples (μg/Kg or μg/L)	Reference
Wheat-based baby foods (25)	Spain/Ireland	-	-	n.d.	[67]
Barley (15)	Czech Republic	-	-	n.d.	[68]
Cereals and cereal-based products (265)	Mediterranean countries	2010	1.5%	FB_1 <LOQ-184 FB_2 121-176	[69]
Cereals (100)	Malaysia	2010	19%	FB_1 41.3-209.3 FB_2 40.1-113.5	[70]
Cereals (80)	Malaysia	2010	16.3%	FB_1 10.75-33.25 FB_2 11.35-31.19	[71]
Wheat (103)	Servia	2005 and 2007	89.3%	FB_1 750-5400	[72]
Beer (33)	European countries	2010-2011	97%	FB_1 <0.1-30.3 FB_2 <0.1-3.9	[73]
Maize (83)	Italy	2006-2007	96% in 2006 94% in 2007	FUM <25-76,323	[74]
Flour (50)	Spain	-	6%	FB_2 230-468	[75]
Maize (178)	Italy	2007	?	FUM 210.79-3662.96	[76]
Maize and maize-based products (27)	Italy	-	100%	-	[77]
Cereals, legumes and nuts (108)	Jordan	2008-2009	0.9%	FUM 250*	[78]
Vine fruit (13)	USA/ Turkey/ China/ India/ Iran/ South Africa	-	61.5%	FB_1 <LOQ-17,321.51 FB_2 <LOQ-11,515.78 FB_3 <LOQ-3716.03 FB_4 <LOQ-1483.92 FB_5 43.80*	[79]

Mean: mean concentration of fumonisins detected in samples.

n.d.: not detected.

FUM: total of fumonisins.

<LOQ: less than the limit of quantification.

*Just one sample positive.

The presence of mycotoxins in food and feed is inevitable, and as such, humans and animals are exposed to them on a continuous basis leading to a wide array of health complications. Foods and feedstuffs can be contaminated with mycotoxins before harvest, during the time between harvesting and drying, and in storage [14, 15].

According Manyes et al. [10], the fumonisin B_1 can be found in beer, rice, sorghum, triticale, cowpea seeds, beans, soybeans, not only in maize and maize-based products, thus the fumonisins can be found in numerous substrates.

Fumonisins are not metabolized by animals, as a consequence, human contamination has two kind of sources, the consumption of directly contaminated maize-based dietary staples or of indirectly toxic food produced by infected animals, so that, their occurrences could be found in milk, cheese, meat or eggs [16].

Fumonisins have a certain thermal stability, maize meal contaminated at concentration level 2.5 µg/g after heating at 190 °C for 60 minutes reduced the fumonisins by over 70%, while heated at 220 °C for 25 minutes this reduction was 100%. Fumonisin contamination decreased 60% in maize meal muffins baked at 220 °C for 25 minutes [12, 16]. This shows that processing can affect the amount of toxin present in foods.

Table 1 shows the occurrence of fumonisins in foods, indicating the presence of the toxin in various food matrices and high variation in concentrations. This variation in contamination levels depend of the substrate and the climatic conditions of the country.

MANAGEMENT PRACTICES: CONTROL OF FUMONISINS LEVEL IN FOOD

Occurrence of mycotoxins can cause economic losses due reduction of quality foods for humans and animals, reduction in animal production due to feed refusal or diseases and significant losses related to human health, but unfortunately, there is still not much information in the literature about the impact of the type of agricultural system on the mycotoxin content [80–82].

Fumonisins are unavoidable contaminants in food and feed chains and their presence needs to be minimized, methods for controlling are usually preventive as good agricultural practice and sufficient drying of crops after harvest [10, 83].

The contamination in maize occurs mainly at preharvest, particularly when grown in warmer regions, or during the early stage of storage, thus, *Fusarium* species are usually considered as field fungi and except under extreme conditions, the concentration of fumonisins does not increase during storage [14, 83].

Under favourable conditions, the *Fusarium verticillioides* causes root, stalk, ear, kernel and seedling rot, resulting in serious production losses in maize [7]. Ear rot is more severe when hot, dry weather occurs [84].

Temperature is a primary determining factor that modulates fungal growth and mycotoxin production, *Fusarium verticillioides* and *Fusarium proliferatum* can grow over a wide range of temperatures, but *Fusarium* growth is more common in temperate weathers at temperatures ranging from 26–28 °C [14, 80].

Water activity (aw) is another important factor, *F. verticillioides* and *F. proliferatum* can grow only at relatively high water activities (a_w> 0.88) [14, 80].

Climatic conditions as drought and pluvial precipitation, insect damage and plant characteristics, play an important role in mycotoxin contamination [80, 85].

The most effective steps of prevention are those carried out before the fungal infestation and before mycotoxin production. Different strategies have been developed to diminish the contamination, but even the best management of agricultural strategies cannot totally eradicate mycotoxin contamination [86].

Fusarium spores can be spread everywhere by wind, and then settle on the soil and stay for a long period or grow on dead plant residues such as straw or stubbles [86]. Cotten and Munkvold [87] assessed the duration that specific strains of *F. verticillioides* and *F. proliferatum* can survive in maize stalk residue under field conditions, after 630 days the *Fusarium* species were recovered and the survival was greatest in the residues that were on the surface.

Greater recovery from surface residues than from buried residues in this experiment suggests that tillage (techniques used for soil cultivation, as ploughing) could reduce inoculum sources for these fungi, but the tillage as well as crop rotation are not always recognized as efficient [86, 87].

Fertilizers can affect the incidence and severity of *Fusarium* sp. by altering the rate of residue decomposition, by acting on the rate of plant growth, by creating a physiological stress on the host plant and by changing the soil structure and its microbial activity [86, 88].

Transgenic Bt maize contains a gene from the *Bacillus thuringiensis* (soil bacterium), which encodes for formation of a crystal (Cry) protein that is toxic to common lepidopteran maize pests, thus, the reduction of pest damage has indirect benefit at lower levels of mycotoxin contamination, but the plant breeding can be considered as the best solution for *Fusarium* control in susceptible crops [86, 89].

Microbial competitors can be used to limit the growth of fungi, Ferrigo et al. [90], determined the efficacy of seed treatment with *Trichoderma harzianum* on *F. verticillioides* maize colonization and on fumonisin contamination under natural conditions during three years, and they found that, this seed treatment reduces *F. verticillioides* infection and fumonisin contamination under field conditions, suggesting as a promising and environmentally friendly way.

The use of synthetic fungicides is common to protect food products from deterioration by fungi and mycotoxin contamination, but is being discouraged due to economic reasons and growing concern for environment and food safety issues [91, 92].

There is an increased interest on the study and utilization of antifungal compounds obtained from natural sources to replace synthetic fungicides, interest has been focused on the potential application of plant essential oils to prevent fungal growth and mycotoxins contamination in the cereals and grain-based food [83, 93].

Many essential oils have been tested for their effects against *Fusarium* species, including *F. verticillioides*and fumonisin production, as the essential oils of *Rosmarinu sofficinalis*, *Pelargonium graveolens*, *Ocimum basilicum*, *Cuminum cyminum*, *Cinnamomum zeylanicum*, *Eugenia caryophyllata*, *Cymbopogon citratus*, *Origanum vulgare*, *Cymbopogon martinii*, *☐ataria multiflora*, *Heracleum persicum*, *Cuminum cyminum*, *Pinaceae*, *Foeniculum vulgare* and *☐ingiber officinale* [94–98].

The processing of maize and the form of the product used as food has an important bearing on the exposure assessment, the fate of fumonisins during various processing stages has been the subject of research papers and published results indicate that reductions in contamination levels can be achieved [99].

Pitt, Taniwaki and Cole [100], presented in their paper a general overview of how the increases and decreases in mycotoxin levels in foods occur during the growing, harvesting, storage and processing.

In the preharvest, *F. verticillioides* is endemic in maize, causes little damage and little fumonisin formation. However, drought stress and insect damage cause a great increase in growth of the fungus, and fumonisin production [100].

During the postharvest, the rapid drying is recommended only in the initial stages of drying, then when the moisture content is reduced the fumonisin production ceases [100].

In storage, *Fusarium* species grow very little, so fumonisins will not be produced, under all normal conditions [100].

In most often in the processing, the main methods for meeting quality are the visual inspection of lots for fungal damage, fumonisin analyses, and rejection of lots that do not meet specifications, some processing has effect on fumonisin concentrations and can reduce the levels [100].

The nixtamalization, for example, is the process that involves cooking and steeping maize in alkaline water and the fumonisin concentrations in the cooked product are reduced during this step by hydrolysis of one or both of the molecule's tricarballylic acid groups, changing the structure of the fumonisin and forming the hydrolyzed fumonisin (HFB) [101, 102].

Burns et al. [101], observed in their paper that nixtamalization effectively reduce the amount of fumonisins and lesions in the kidney of rats, when compared to a diet with FB_1, concluding that the nixtamalization reduce the bioavailability and toxicity of FB_1.

Grenier et al. [103] assessed in their study the effects on the liver and intestines of piglets exposed to 2.8 μmol of FB_1 or hydrolyzed fumonisin FB_1 (HFB_1)/kg b.w./day for 14 days. Signs of hepatotoxicity were observed upon ingestion of FB_1 and the intestine of animals exposed to this mycotoxin displayed mild to moderate tissular lesions including villi atrophy and fusion. The hepatotoxicity was not observed in HFB_1-treated piglets and the intestinal integrity has not altered.

Generotti et al. [6] estimated the distribution of the fumonisin contamination in the various different production chain maize derivatives (unprocessed and cleaned maize, uncooked and cooked broken maize, cornflake and cooled cornflake, flour "fumetto", cornmeal semolina, germ and middling) collected from a cornmeal industrial production plant, to study the influence of industrial processing. The results showed that fumonisin amounts decreased significantly after cleaning and corn flaking processes, achieving a reduction of about 40% in the cornmeal semolina, while higher level of contamination in maize flour (fumetto) and middling were detected.

Fumonisin contamination is practically inevitable, so several countries have implemented laws that limit the maximal levels by this toxin. The Food and Drug Administration (FDA) recommended maximum levels for fumonisins in human foods which ranging, to $FB_1+FB_2+FB_3$, 2000-4000 µg/Kg, in degermed dry milled corn products, dry milled corn bran, cleaned corn intended for masa production, cleaned corn intended for popcorn, among others [104]. Food and Agriculture Organization of the United Nations (FAO), also recommends maximum levels for fumonisins in maize, ranging from 1000-3000 µg/kg [105]. These limits are a strategy for the control of fumonisins in food, in order that there is not fully effective method for its total elimination.

HEALTH CONCERNS

Despite almost 30 years of research, the mechanism of toxicity of fumonisinsis not yet fully understood, due to its complexity [8, 13]. The FB_1 is structurally similar to sphingoid bases such as sphingosine, which is a component of the sphingolipid molecule, and in high concentrations FB_1 inhibits ceramide synthase, which is an important enzyme in the *de novo* pathway of sphingolipid biosynthesis, leading at changes of the ratio sphinganine/sphingosine (Sa/So) [13, 106].

Fumonisins has a free amino group at C-2 and through noncovalent interactions inhibit the ceramide synthase [106]. The first microscopic evidence of exposure of FBs and ceramide synthase inhibition is apoptosis, mitosis and Sa and So accumulation in liver and kidney cells, but the inactivation of this enzyme can promote too, an increases free Sa and So, sphingosine 1-phosphate (So 1-P) and decreases complex sphingolipids such as ceramides, sphingomyelin, ganglioside and glycosphingolipids in tissues, blood and urine, as well as, impairs metabolism of arachidonic acid and reduces re-acylation of So [13].

Various aspects of cellular regulation are linked to the sphingolipids and a sphingolipid metabolism disorder may form a suitable groundwork for toxicity, fumonisins can impair complex sphingolipids that are involved in synthesis and transport, block sphingolipid biosynthesis and lead to degeneration of the sphingolipid-rich tissues [13].

According to Chuturgoon, Phulukdaree and Moodley [107], although the cytotoxic mechanism of FB_1 is attributed to the disruption of sphingolipid metabolism, the underlying mechanisms of its cancer initiating properties are unknown, therefore, disruption of sphingolipid metabolism may not fully explain the proliferative and oncogenic properties of this mycotoxin, thus, in their report, the authors investigated the effect of FB_1 on enzymes, DNA methyltransferases and demethylases, involved in chromatin maintenance and gross changes in structural integrity of DNA in HepG2 cells.

The study showed that FB_1 significantly decreased the expression of DNA methyltransferases and concomitantly significantly increased expression of demethylases and the histone demethylases. In conclusion, the results showed that FB_1 upsets the balance of enzymes necessary for proper regulation of DNA methylation in human hepatoma cells that causes chromatin instability and may lead to liver tumorigenesis [107].

International Agency for Research on Cancer classified fumonisin B_1 in group 2B, i.e. possibly carcinogenic to humans [108].

The presence of fumonisin B_1 in maize grains has been associated with cases of equine leukoencephalomalacia and porcine pulmonary edema [10, 92, 109, 110]; human esophageal cancer [92]; neural tube defects in newborns [13]; cancer in experimental animals [13].

Equidae and porcine species are considered to be the most sensitive animal species to fumonisins, developing specific clinical syndromes, for example, the symptoms of equine leukoencephalomalacia include ataxia, paresis, apathy, hypersensitivity, impaired locomotor function, necrosis of cerebral white matter, and lesions in the cerebral [92, 110].

Several studies have correlated the high incidence of contamination by fumonisin with the occurrence of esophageal cancer, in high esophageal cancer prevalence areas the exposure to fumonisins was higher in studies in China, Southern Africa, Iran and Brazil [111–114].

Fumonisin B_1 reduces uptake of folate in different cell lines, so their consumption has been implicated in neural tube defects in human babies [92]. In a study, pregnant mice were injected with increasing doses of FB_1 and their exposed fetuses were examined for malformations, exposure altered sphingolipid metabolism and folate concentrations, 79% of the exposed fetuses demonstrated neural tube defects [115]

Howard et al. [116] examined the carcinogenicity of FB_1 in rodents. Their results showed that FB_1 induces renal tubule tumors in male F344 rats and hepatic tumors in female B6C3F1 mice, when included in the diets.

Souza et al. [8], analyzed male Wistar developing rats fed with diets that contained fumonisins B_1+B_2 at concentrations of 1 and 3 mg/kg to evaluate the effects of fumonisins on the myenteric plexus of the jejunum, the fumonisins concentrations used are equivalent to food preparations that contain corn in a proportion of 34%. The results indicated that a fumonisin-containing diet affected the growth process of myenteric neurons, indicating that fumonisin intake negatively influences neuroplastic mechanisms in the small intestine in rats between 21 and 63 days of age corroborating with several studies of the effect of these mycotoxins on the development of neurons in the central nervous system.

CONCLUSION

In mild climates, the occurrence of fusaria and their toxins are unavoidable contaminants in food and feed chains. The presence of fumonisins can affect various food matrices and represents a significant hazard to human health. The control of fumonisin contamination is very difficult, therefore it is essential the good agricultural practices, particularly at preharvest, when the major contamination in maize occurs.

Fungal growth and mycotoxin production result from a complex interaction between many factors and understanding it is essential to known the overwall process and to prevent mycotoxin production. An inevitable part of the preventive measures is regular foodstuffs monitoring with mycological and mycotoxicological examinations.

REFERENCES

[1]   Čonková, E. Laciaková, A. Kováč, G. & Seidel, H. (2003). Fusarial toxins and their role in animal diseases. *The veterinary Journal, 165,* 214–220.

[2]   Pitt, J. I. & Hocking, A. D. (1997). *Fungi and Food Spoilage* (2 ed). London, United Kingdom: Blackie Academic &Professional.

[3]   Burgess, L. W. Summerell, B. A. Bullock, S. Gott, K. P. & Backhouse, D. (1994). *Laboratory manual for fusarium research* (1 ed). Sydney, Australia: University of Sydney.

[4]   Marasas, W. F. O. Nelson, P. E. & Toussoun, T. A. (1984). *Toxigenic Fusarium species: identity and mycotoxicology* (1 ed). Washington, United States: Pennsylvania State University.

[5]   Creppy, E. E. (2002). Update of survey, regulation and toxic effects of mycotoxins in Europe.*Toxicology Letters, 127,*19–28.

[6]   Generotti, S. Cirlini, M. Dall'Asta, C. & Suman, M. (2015). Influence of the industrial process from caryopsis to cornmeal semolina on levels of fumonisins and their masked forms. *Food Control, 48,* 170–174.

[7] Menniti, A. M. Gregori, R. & Neri, F. (2010). Activity of natural compounds on *Fusariumverticillioides* and fumonisin production in stored maize kernels. *International Journal of Food Microbiology, 136*, 304–309.

[8] Sousa, F. C. Schamber, C. R. Amorin S. S. S. & Natali, M. R. M. (2014). Effect of fumonisin-containing diet on the myenteric plexus of the jejunum in rats. *Autonomic Neuroscience: Basic and Clinical, 185*, 93–99.

[9] Wan Norhasima, W. M. Abdulamir, A. S. Abu Bakar, F. Son, R. & Norhafniza, A. (2009). The health and toxic adverse effects of *Fusarium* fungal mycotoxin , fumonisins , on human population. *American Journal of Infectious Diseases,5*, 283–291.

[10] Manyes, L. Ruiz, M. J. Luciano, F. B. & Meca, G. (2014). Bioaccessibility and bioavailability of fumonisin B_2 and its reaction products with isothiocyanates through a simulated gastrointestinal digestion system. *Food Control, 37*, 326–335.

[11] Rocha, M. E. B. Freire, F. C. O. Maia, F. E. F. Guedes, M. I. F. & Rondina, D. (2014). Mycotoxins and their effects on human and animal health. *Food Control, 36*, 159–165.

[12] Waśkiewicz, A. Beszterda, M. & Goliński, P. (2012). Occurrence of fumonisins in food - An interdisciplinary approach to the problem. *Food Control, 26*, 491–499.

[13] Escrivá, L. Font, G. & Manyes, L. (2015). *In vivo* toxicity studies of *Fusarium* mycotoxins in the last decade: A review. *Food Chemical Toxicology, 78*, 185–206.

[14] Marin, S. Ramos, A. J. Cano-Sancho, G. & Sanchis, V. (2013). Mycotoxins: Occurrence, toxicology, and exposure assessment. *Food Chemical Toxicology, 60*, 218–237.

[15] Cast, Council for Agricultural Science and Technology. *Mycotoxins: Risks in plant, animal, and human systems.* (2003). Available from: http://www.trilogylab.com/uploads/Mycotoxin_CAST_Report.pdf

[16] Giacomo, D. R. & Stefania, D. Z. (2013). A multivariate regression model for detection of fumonisins content in maize from near infrared spectra. *Food Chemistry, 141*, 4289–4294.

[17] Li, R. Tao, B. Pang, M. Liu, Y. & Dong, J. (2015). Natural occurrence of fumonisins B_1 and B_2 in maize from three main maize-producing provinces in China. *Food Control, 50*, 838–842.

[18] Kirinčič, S. Škrjanc, B. Kos, N. Kozolc, B. Pirnat, N. & Tavčar-Kalcher, G. (2015). Mycotoxins in cereals and cereal products in Slovenia –

Official control of foods in the years 2008–2012. *Food Control, 50*, 157–165.

[19] Bolechová, M. Benešová, K. Běláková, S. Čáslavský, J. Pospíchalová, M. & Mikulíková, R. (2015). Determination of seventeen mycotoxins in barley and malt in the Czech Republic. *Food Control, 47*, 108–113.

[20] Bordin, K. Rottinghaus, G. E. Landers, B. R. Ledoux, D. R. Kobashigawa, E. Corassin, C. H. & Oliveira, C. A. F. (2015). Evaluation of fumonisin exposure by determination of fumonisin B_1 in human hair and in Brazilian corn products. *Food Control, 53*, 67–71.

[21] Ediage, E. N. Poucke, C. V. & Saeger, S. (2015). A multi-analyte LC–MS/MS method for the analysis of 23 mycotoxins in different sorghum varieties: The forgotten sample matrix. *Food Chemistry, 177*, 397–404.

[22] Gazzotti, T. Biagi, G. Pagliuca, G. Pinna, C. Scardilli, M. Grandi, M. & Zaghini, G. (2015). Occurrence of mycotoxins in extruded commercial dog food. *Animal Feed Science and Technology, 202*, 81–89.

[23] Gutleb, A. C. Caloni, F. Giraud, F. Cortinovis, C. Pizzo, F. Hoffmann, L. Bohn, T. & Pasquali, M. (2015). Detection of multiple mycotoxin occurrences in soy animal feed by traditional mycological identification combined with molecular species identification. *Toxicology Reports, 2*, 275–279.

[24] Kamala, A. Ortiz, J. Kimanya, M. Haesaert, G. Donoso, S. Tiisekwa, B. & Meulenaer, B. (2015). Multiple mycotoxin co-occurrence in maize grown in three agro-ecological zones of Tanzania. *Food Control, 54*, 208–215.

[25] Mata, A. T. Ferreira, J. P. Oliveira, B. R. Batoréu, M. C. Barreto Crespo, M. T. Pereira, V. J. & Bronze, M. R. (2015). Bottled water: Analysis of mycotoxins by LC–MS/MS. *Food Chemistry, 176*, 455–464.

[26] Piacentini, K. C. Savi, G.D. Olivo, G. & Scussel, V. M. (2015). Quality and occurrence of deoxynivalenol and fumonisins in craft beer. *Food Control, 50*, 925–929.

[27] Cendoya, E. Monge, M. P. Palacios, S. A. Chiacchiera, S. M. Torres, A. M. Farnochi, M. C. & Ramirez, M. L. (2014). Fumonisin occurrence in naturally contaminated wheat grain harvested in Argentina. *Food Control, 37*, 56–61.

[28] Kos, J. Hajnal, E. J. Škrinjar, M. Mišan, A. Mandić, A. Jovanov, P. & Milovanović, I. (2014). Presence of *Fusarium* toxins in maize from Autonomous Province of Vojvodina, Serbia. *Food Control, 46*, 98–101.

[29] Magoha, H. Meulenaer, B. Kimanya, M. Hipolite, D. Lachat, C. & Kolsteren, P. (2014). Fumonisin B_1 contamination in breast milk and its

exposure in infants under 6 months of age in Rombo, Northern Tanzania. *Food Chemical Toxicology, 74*, 112–116.

[30] Matumba, L. Poucke, C. V. Biswick, T. Monjerezi, M. Mwatseteza, J. & Saeger, S. (2014). A limited survey of mycotoxins in traditional maize based opaque beers in Malawi. *Food Control, 36*, 253–256.

[31] Petrarca, M. H. Rodrigues, M. I. Rossi, E. A. & Sylos, C. M. (2014). Optimisation of a sample preparation method for the determination of fumonisin B$_1$ in rice. *Food Chemistry, 158*, 270–277.

[32] Azaiez, I. Font, G. Mañes, J. & Fernández-Franzón, M. (2015). Survey of mycotoxins in dates and dried fruits from Tunisian and Spanish markets. *Food Control, 51*, 340–346.

[33] Alkadri, D. Rubert, J. Prodi, A. Pisi, A. Mañes, J. & Soler, C. (2014). Natural co-occurrence of mycotoxins in wheat grains from Italy and Syria. *Food Chemistry, 157*, 111–118.

[34] Arroyo-Manzanares, N. Huertas-Pérez, J. F. García-Campaña, A.M. & Gámiz-Gracia, L. (2014). Simple methodology for the determination of mycotoxins in pseudocereals, spelt and rice. *Food Control, 36*, 94–101.

[35] Blesa, J. Moltó, J. C. Akhdari, S. E. Mañes, J. & Zinedine, A. (2014). Simultaneous determination of *Fusarium* mycotoxins in wheat grain from Morocco by liquid chromatography coupled to triple quadrupole mass spectrometry. *Food Control, 46*, 1–5.

[36] Bordin, K. Rosim, R. E. Neeff, D. V. Rottinghaus, G. E. & Oliveira, C. A. F. (2014). Assessment of dietary intake of fumonisin B$_1$ in São Paulo, Brazil. *Food Chemistry, 155*, 174–178.

[37] Brera, C. Debegnach, F. De Santis, B. Di Ianni, S. Gregori, E. Neuhold, S. & Valitutti, F. (2014). Exposure assessment to mycotoxins in gluten-free diet for celiac patients. *Food and Chemical Toxicology, 69*, 13–17.

[38] Chala, A. Taye, W. Ayalew, A. Krska, R. Sulyok, M. & Logrieco, A. (2014). Multimycotoxin analysis of sorghum (*Sorghum bicolor* L. Moench) and finger millet (*Eleusine coracana* L. Garten) from Ethiopia. *Food Control, 45*, 29–35.

[39] Pizzutti, I. R. Kok, A. Scholten, J. Righi, L. W. Cardoso, C. D. Rohers, G. N. & Silva, R. C. (2014). Development, optimization and validation of a multimethod for the determination of 36 mycotoxins in wines by liquid chromatography-tandem mass spectrometry. *Talanta, 129*, 352–363.

[40] Kimanya, M. E. Shirima, C. P. Magoha, H. Shewiyo, D. H. Meulenaer, B. Kolsteren, P. & Gong, Y. Y. (2014). Co-exposures of aflatoxins with

deoxynivalenol and fumonisins from maize based complementary foods in Rombo, Northern Tanzania. *Food Control, 41*, 76–81.

[41] Yogendrarajah, P. Jacxsens, L. Saeger, S. & Meulenaer, B. (2014). Co-occurrence of multiple mycotoxins in dry chilli (*Capsicum annum* L.) samples from the markets of Sri Lanka and Belgium. *Food Control, 46*, 26–34.

[42] Yogendrarajah, P. Deschuyffeleer, N. Jacxsens, L. Sneyers, P. J. Maene, P. Saeger, S. Devlieghere, F. & Meulenaer, B. (2014). Mycological quality and mycotoxin contamination of Sri Lankan peppers (*Piper nigrum* L.) and subsequent exposure assessment. *Food Control, 41*, 219–230.

[43] Zachariasova, M. Dzuman, Z. Veprikova, Z. Hajkova, K. Jiru, M. Vaclavikova, M. Zachariasova, A. Pospichalova, M. Florian, M. & Hajslova, J. (2014). Occurrence of multiple mycotoxins in european feedingstuffs, assessment of dietary intake by farm animals. *Animal Feed Science and Technology, 193*, 124–140.

[44] Abia, W. A. Warth, B. Sulyok, M. Krska, R. Tchana, A. N. Njobeh, P. B. Dutton, M. F. & Moundipa, P. F. (2013). Determination of multi-mycotoxin occurrence in cereals, nuts and their products in Cameroon by liquid chromatography tandem mass spectrometry (LC-MS/MS). *Food Control, 31*, 438–453.

[45] Beltrán, E. Ibáñez, M. Portolés, T. Ripollés, C. Sancho, J. V. Yusà, V. Marín, S. & Hernández, F. (2013). Development of sensitive and rapid analytical methodology for food analysis of 18 mycotoxins included in a total diet study. *Analytica Chimica Acta, 783*, 39–48.

[46] Rubert, J. Soriano, J. M. Mañes, J. & Soler, C. (2013). Occurrence of fumonisins in organic and conventional cereal-based products commercialized in France, Germany and Spain. *Food and Chemical Toxicology, 56*, 387–391.

[47] Arroyo-Manzanares, N. García-Campaña, A. M. & Gámiz-Gracia, L. (2013). Multiclass mycotoxin analysis in *Silybum marianum* by ultra high performance liquid chromatography-tandem mass spectrometry using a procedure based on QuEChERS and dispersive liquid-liquid microextraction. *Journal of Chromatography A, 1282*, 11–19.

[48] Escobar, J. Lorán, S. Giménez, I. Ferruz, E. Herrera, M. Herrera, A. & Ariño, A. (2013). Occurrence and exposure assessment of *Fusarium* mycotoxins in maize germ, refined corn oil and margarine. *Food and Chemical Toxicology, 62*, 514–520.

[49] Becker-Algeri, T. A. Heidtmann-Bemvenuti, R. Hackbart, H. C. S. & Badiale-Furlong, E. (2013). Thermal treatments and their effects on the fumonisin B_1 level in rice. *Food Control, 34*, 488–493.

[50] Haas, D. Pfeifer, B. Reiterich, C. Partenheimer, R. Reck, B. & Buzina, W. (2013). Identification and quantification of fungi and mycotoxins from Pu-erh tea. *International Journal of Food Microbiology, 166*, 316–322.

[51] Kuang, Y. Qiu, F. Kong, W. Luo, J. Cheng, H. & Yang, M. (2013). Simultaneous quantification of mycotoxins and pesticide residues in ginseng with one-step extraction using ultra-high performance liquid chromatography-electrospray ionization tandem mass spectrometry. *Journal of Chromatography B, 939*, 98–107.

[52] Pleadin, J. Vahčić, N. Perši, N. Ševelj, D. Markov, K. & Frece, J. (2013). *Fusarium* mycotoxins' occurrence in cereals harvested from Croatian fields. *Food Control, 32*, 49–54.

[53] Rubert, J. Fapohunda, S. O. Soler, C. Ezekiel, C.N. Mañes, J. & Kayode, F. (2013). A survey of mycotoxins in random street-vended snacks from Lagos, Nigeria, using QuEChERS-HPLC-MS/MS. *Food Control, 32*, 673–677.

[54] Waśkiewicz, A. Beszterda, M. Bocianowski, J. & Goliński, P. (2013). Natural occurrence of fumonisins and ochratoxin A in some herbs and spices commercialized in Poland analyzed by UPLC-MS/MS method. *Food Microbiology, 36*, 426–431.

[55] Somorin, Y. M. Bertuzzi, T. Battilani, P. & Pietri, A. (2012). Aflatoxin and fumonisin contamination of yam flour from markets in Nigeria. *Food Control, 25*, 53–58.

[56] Gaspardo, B. Del Zotto, S. Torelli, E. Cividino, S. R. Firrao, G. Della Riccia, G. & Stefanon, B. (2012). A rapid method for detection of fumonisins B_1 and B_2 in corn meal using Fourier transform near infrared (FT-NIR) spectroscopy implemented with integrating sphere. *Food Chemistry, 135*, 1608–1612.

[57] Garrido, C. E. Pezzani, C. H. & Pacin, A. (2012). Mycotoxins occurrence in Argentina's maize (*Zea mays* L.), from 1999 to 2010. *Food Control, 25*, 660–665.

[58] Borutova, R. Aragon, Y. A. Nährer, K. & Berthiller, F. (2012). Co-occurrence and statistical correlations between mycotoxins in feedstuffs collected in the Asia-Oceania in 2010. *Animal Feed Science Technology, 178*, 190–197.

[59] Cano-Sancho, G. Ramos, A. J. Marín, S. & Sanchis, V. (2012). Presence and co-occurrence of aflatoxins, deoxynivalenol, fumonisins and zearalenone in gluten-free and ethnic foods. *Food Control, 26*, 282–286.

[60] Škrbić, B. Živančev, J. Durišić-Mladenović, N. & Godula, M. (2012). Principal mycotoxins in wheat flour from the Serbian market: Levels and assessment of the exposure by wheat-based products. *Food Control, 25*, 389–396.

[61] Lattanzio, V. M.T. Nivarlet, N. Lippolis, V. Gatta, S. D. Huet, A. C. Delahaut, P. Granier, B. & Visconti, A. (2012). Multiplex dipstick immunoassay for semi-quantitative determination of *Fusarium* mycotoxins in cereals. *Analytica Chimica Acta, 718*, 99–108.

[62] Martins, F. A. Ferreira, F. M. D. Ferreira, F. D. Bando, É. Nerilo, S.B. Hirooka, E. Y. & Machinski Jr, M. (2012). Daily intake estimates of fumonisins in corn-based food products in the population of Parana, Brazil. *Food Control, 26*, 614–618.

[63] Oueslati, S. Romero-González, R. Lasram, S. Frenich, A. G. & Vidal, J. L.M. (2012). Multi-mycotoxin determination in cereals and derived products marketed in Tunisia using ultra-high performance liquid chromatography coupled to triple quadrupole mass spectrometry. *Food and Chemical Toxicology, 50*, 2376–2381.

[64] Queiroz, V. A.V. Alves, G. L. O. Conceição, R. R.P. Guimarães, L. J. M. Mendes, S. M. Ribeiro, P. E. A. & Costa, R. V. (2012). Occurrence of fumonisins and zearalenone in maize stored in family farm in Minas Gerais, Brazil. *Food Control, 28*, 83–86.

[65] Rubert, J. Soler, C. & Mañes, J. (2012). Application of an HPLC-MS/MS method for mycotoxin analysis in commercial baby foods. *Food Chemistry, 133*, 176–183.

[66] Rubert, J. Soler, C. & Mañes, J. (2012). Occurrence of fourteen mycotoxins in tiger-nuts. *Food Control, 25*, 374–379.

[67] Rubert, J. James, K. J. Mañes, J. & Soler, C. (2012). Applicability of hybrid linear ion trap-high resolution mass spectrometry and quadrupole-linear ion trap-mass spectrometry for mycotoxin analysis in baby food. *Journal of Chromatography A, 1223*, 84–92.

[68] Rubert, J. Dzuman, Z. Vaclavikova, M. Zachariasova, M. Soler, C. & Hajslova, J. (2012). Analysis of mycotoxins in barley using ultra high liquid chromatography high resolution mass spectrometry: Comparison of efficiency and efficacy of different extraction procedures. *Talanta, 99*, 712–719.

[69] Serrano, A. B. Font, G. Ruiz, M. J. & Ferrer, E. (2012). Co-occurrence and risk assessment of mycotoxins in food and diet from Mediterranean area. *Food Chemistry, 135*, 423–429.

[70] Soleimany, F. Jinap, S. & Abas, F. (2012). Determination of mycotoxins in cereals by liquid chromatography tandem mass spectrometry. *Food Chemistry, 130*, 1055–1060.

[71] Soleimany, F. Jinap, S. Faridah, A. & Khatib, A. (2012). A UPLC-MS/MS for simultaneous determination of aflatoxins, ochratoxin A, zearalenone, DON, fumonisins, T-2 toxin and HT-2 toxin, in cereals. *Food Control, 25*, 647–653.

[72] Stanković, S. Lević, J. Ivanović, D. Krnjaja, V. Stanković, G. & Tančić, S. (2012). Fumonisin B_1 and its co-occurrence with other fusariotoxins in naturally-contaminated wheat grain. *Food Control, 23*, 384–388.

[73] Bertuzzi, T. Rastelli, S. Mulazzi, A. Donadini, G. & Pietri, A. (2011). Mycotoxin occurrence in beer produced in several European countries. *Food Control, 22*, 2059–2064.

[74] Covarelli, L. Beccari, G. & Salvi, S. (2011). Infection by mycotoxigenic fungal species and mycotoxin contamination of maize grain in Umbria, central Italy. *Food and Chemical Toxicology, 49*, 2365–2369.

[75] Rubert, J. Soler, C. & Mañes, J. (2011). Evaluation of matrix solid-phase dispersion (MSPD) extraction for multi-mycotoxin determination in different flours using LC-MS/MS. *Talanta, 85*, 206–215.

[76] Firrao, G. Torelli, E. Gobbi, E. Raranciuc, S. Bianchi, G. & Locci, R. (2010). Prediction of milled maize fumonisin contamination by multispectral image analysis. *Journal of Cereal Science, 52*, 327–330.

[77] Anfossi, L. Calderara, M. Baggiani, C. Giovannoli, C. Arletti, E. & Giraudi, G. (2010). Development and application of a quantitative lateral flow immunoassay for fumonisins in maize. *Analytica Chimica Acta, 682*, 104–109.

[78] Salem, N. M. & Ahmad, R. (2010). Mycotoxins in food from Jordan: Preliminary survey. *Food Control, 21*, 1099–1103.

[79] Varga, J. Kocsubé, S. Suri, K. Szigeti, G. Szekeres, A. Varga, M. Tóth, B. & Bartók, T. (2010). Fumonisin contamination and fumonisin producing blackAspergilli in dried vine fruits of different origin. *International Journal of Food Microbiology, 143*, 143–149.

[80] Marroquín-Cardona, A. G. Johnson, N. M. Phillips, T. D. & Hayes, A. W. (2014). Mycotoxins in a changing global environment - A review. *Food and Chemical Toxicology, 69*, 220–230.

[81] Ahmad, B. Ashiq, S. Hussain, A. Bashir, S. & Hussain, M. (2014). Evaluation of mycotoxins, mycobiota, and toxigenic fungi in selected medicinal plants of Khyber Pakhtunkhwa, Pakistan. *Fungal Biology, 118*, 776–784.

[82] Błajet-Kosicka, A. Twaruzek, M. Kosicki, R. Sibiorowska, E. & Grajewski, J. (2014). Co-occurrence and evaluation of mycotoxins in organic and conventional rye grain and products. *Food Control, 38*, 61–66.

[83] Ferrochio, L. V. Cendoya, E. Zachetti, V. G. L. Farnochi, M. C. Massad, W. & Ramirez, M. L. (2014). Combined effect of chitosan and water activity on growth and fumonisin production by *Fusarium verticillioides* and *Fusarium proliferatum* on maize-based media. *International Journal of Food Microbiology, 185*, 51–56.

[84] Lanubile, A. Passini, L. & Marocco, A. M. (2010). Differential gene expression in kernels and silks of maize lines with contrasting levels of ear rot resistance after *Fusarium verticillioides* infection. *Journal of Plant Physiology, 167*, 1398–1406.

[85] Cao, A. Santiago, R. Ramos, A. J. Souto, X. C. Aguín, O. Malvar, R. A. & Butrón, A. (2014). Critical environmental and genotypic factors for *Fusarium verticillioides* infection, fungal growth and fumonisin contamination in maize grown in northwestern Spain. *International Journal of Food Microbiology, 177*, 63–71.

[86] Jouany, J. P. (2007). Methods for preventing , decontaminating and minimizing the toxicity of mycotoxins in feeds. *Animal Feed Science and Tecnology, 137*, 342–362.

[87] Cotton, T. K. & Munkvold, G. P. (1998). Survival of *Fusarium moniliforme , F . proliferatum ,* and *F . subglutinans* in Maize Stalk Residue. *Phytopathology, 88*, 550–555.

[88] Edwards, S. G. (2004). Influence of agricultural practices on *Fusarium* infection of cereals and subsequent contamination of grain by trichothecene mycotoxins. *Toxicology, 153*, 29–35.

[89] Wu, F. (2006). Mycotoxin reduction in Bt corn: potential economic , health , and regulatory impacts. *Transgenic Research,* 15, 277–289.

[90] Ferrigo, D. Raiola, A. Rasera, R. & Causin, R. (2014). *Trichoderma harzianum* seed treatment controls *Fusarium verticillioides* colonization and fumonisin contamination in maize under field conditions. *Crop Protection, 65*, 51–56.

[91] Dambolena, J. S. Zygadlo, J. A. & Rubinstein, H. R. (2011). Antifumonisin activity of natural phenolic compounds. A structure-

property-activity relationship study. *International Journal of Food Microbiology, 145*, 140–146.

[92] Zain, M. E. (2011). Impact of mycotoxins on humans and animals. *Journal of Saudi Chemical Society, 15*, 129–144.

[93] Dambolena, J. S. Zunino, M. P. López, A.G. Rubinstein, H.R. Zygadlo, J. A. Mwangi, J. W. Thoithi, G. N. Kibwage, I. O. Mwalukumbi, J. M. & Kariuki, S. T. (2010). Essential oils composition of *Ocimum basilicum* L. and *Ocimum gratissimum* L. from Kenya and their inhibitory effects on growth and fumonisin production by *Fusarium verticillioides.Innovative Food Science and Emerging Technologies, 11*, 410–414.

[94] Bomfim, N. S. Nakassugi, L. P. Oliveira, F. P. J. Kohiyama, C. Y. Mossini, S. A. G. Grespan, R. Nerilo, S. B. Mallmann, C. A. Filho, B. A. A. & Machinski Jr, M. (2015). Antifungal activity and inhibition of fumonisin production by *Rosmarinus officinalis* L. essential oil in *Fusarium verticillioides* (Sacc.) Nirenberg. *Food Chemistry, 166*, 330–336.

[95] Hashem, M. Moharam, A. M. Zaied, A. A. & Saleh, F. E. M. (2010). Efficacy of essential oils in the control of cumin root rot disease caused by *Fusarium* spp. *Crop Protection, 29*, 1111–1117.

[96] Marín, S. Velluti, A. Ramos, A. J. & Sanchis, V. (2004). Effect of essential oils on zearalenone and deoxynivalenol production by *Fusarium graminearum* in non-sterilized maize grain. *Food Microbiology, 21*, 313–318.

[97] Naeini, A. Ziglari, T. Shokri, H. & Khosravi, A. R. (2010). Assessment of growth-inhibiting effect of some plant essential oils on different *Fusarium* isolates. *Journal de Mycologie Médicale, 20*, 174–178.

[98] Yamamoto-Ribeiro, M. M. G. Grespan, R. Kohiyama, C. Y. Ferreira, F. D. Mossini, S. A. G. Silva, E. L. Filho, B. A. A. Mikcha, J. M. G. & Machinski Jr, M. (2013). Effect of ⬜*ingiber officinale* essential oil on *Fusarium verticillioides* and fumonisin production. *Food Chemistry, 141*, 3147–3152.

[99] Shephard, G. S. Kimanya, M. E. Kpodo, K. A. Gnonlonfin, G. J. B. & Gelderblom, W. C. A. (2013). The risk management dilemma for fumonisin mycotoxins. *Food Control, 34*, 596–600.

[100] Pitt, J. I. Taniwaki, M. H. & Cole, M. B. (2013). Mycotoxin production in major crops as influenced by growing, harvesting, storage and processing, with emphasis on the achievement of Food Safety Objectives. *Food Control, 32*, 205–215.

[101] Burns, T.D. Snook, M. E. Riley, R. T. & Voss, K. A. (2008). Fumonisin concentrations and *in vivo* toxicity of nixtamalized *Fusarium verticillioides* culture material: Evidence for fumonisin-matrix interactions. *Food and Chemical Toxicology, 46*, 2841–2848.

[102] Gazzotti, T. Zironi, E. Lugoboni, B. Barbarossa, A. Piva, A. & Pagliuca, G. (2011). Analysis of fumonisins B_1, B_2 and their hydrolysed metabolites in pig liver by LC-MS/MS. *Food Chemistry, 125*, 1379–1384.

[103] Grenier, B. Bracarense, A. P. F.L. Schwartz, H. E. Trumel, C. Cossalter, A. M. Schatzmayr, G. Kolf-Clauw, M. Moll, W. D. & Oswald, I. P. (2012). The low intestinal and hepatic toxicity of hydrolyzed fumonisin B_1 correlates with its inability to alter the metabolism of sphingolipids. *Biochemical Pharmacology, 83*, 1465–1473.

[104] FDA, Food and Drug Administration. Fumonisin Levels in Human Foods and Animal Feeds. (2001). Available from: http://www.fda.gov/Food/GuidanceRegulation/Guidanc.

[105] FAO, Food and Agriculture Organization of the United Nations. Worldwide regulations for mycotoxins in food and feed in 2003. (2004). Available from: http://www.fao.org/docrep/007/y5499e/y5499e07.htm#bm07.4.

[106] Soriano, J. M. González, L. & Catalá, A. I. (2005). Mechanism of action of sphingolipids and their metabolites in the toxicity of fumonisin B_1. *Progress in Lipid Research, 44*, 345–356.

[107] Chuturgoon, A. Phulukdaree, A. & Moodley, D. (2014). Fumonisin B_1 induces global DNA hypomethylation in HepG2 cells - An alternative mechanism of action. *Toxicology, 315*, 65–69.

[108] IARC, International Agency for Research on Cancer. Some traditional herbal medicines, some mycotoxins, naphthalene and styrene. (2002). Available from: http://monographs.iarc.fr/ENG/Monographs/vol82/.

[109] Minervini, F. Debellis, L. Garbetta, A. Girolamo, A. Schena, R. Portincasa, P. & Visconti, A. (2014). Influence on functional parameters of intestinal tract induced by short-term exposure to fumonisins contaminated corn chyme samples. *Food and Chemical Toxicology, 66*, 166–172.

[110] Verstraete, F, (2013), Risk management of undesirable substances in feed following updated risk assessments. *Toxicology and Applied Pharmacology, 270*, 230–247.

[111] Ueno, Y. Iijima, K. Wang, S. D. Sugiura, Y. Sekijima, M. Tanaka, T. Chen, C. & Yu, S. Z. (1997). Fumonisins as a possible contributory risk

factor for primary liver cancer⬚: a 3-year study of corn harvested in in Haimen , China by HPLC and ELISA. *Food and Chemical Toxicology, 35*, 1143–1150.

[112] Sydenham, E. W. Thiel, P. G. Marasas, W. F. O. Shephard, G. S. Schalkwyk, D. J. V. & Koch, K. R. (1990). Natural occurrence of some *Fusarium* mycotoxins im corn from low and high esophageal cancer prevalence areas of Transkei, South Africa. *Journal of Agricultural and Food Chemistry,* 38, 1900–1903.

[113] Shephard, G. S. Marasas, W. F. O. Leggott, N. L. & Yazdanpanah, H. (2000). Natural occurrence of fumonisins in corn from Iran. *Journal of Agricultural and Food Chemistry*, 48, 1860−1864.

[114] Westhuizen, L. V. D. Shephard, G. S. Scussel, V, M, Costa, L. L. F. Vismer, H. F. Rheeder, J. P. & Marasas, W. F. O. (2003). Fumonisin contamination and *Fusarium* incidence in corn from Santa Catarina , Brazil. *Journal of Agricultural and Food Chemistry, 51*, 5574–5578.

[115] Waes, J. G. Starr, L. Maddox, J. Aleman, F. Voss, K. A. Wilberding, J. & Riley, R. T. (2005). Maternal fumonisin exposure and risk for neural tube defects⬚: Mechanisms in an *in vivo* mouse model. *Birth Defects Research (Part A): Clinical and Molecular Teratology, 497*, 487–497.

[116] Howard, P. C. Eppley, R. M. Stack, M. E. Warbritton, A. Voss, K. A. Lorentzen, R. J. Kovach, R. M. & Bucci, T. J. (2001). Fumonisin B_1 carcinogenicity in a two-year feeding study using F344 rats and B6C3F 1 mice. *Environmental Health Perspectives, 109,* 277–282.

In: Fumonisins
Editor: Craig M. Evans

ISBN: 978-1-63482-789-8
© 2015 Nova Science Publishers, Inc.

Chapter 2

FUMONISIN B_1: A POTENTIAL RISK FACTOR FOR NEURAL TUBE DEFECTS

Ana-Marija Domijan, Maja Šegvić Klarić,*
Tihana Žanić Grubišić and Lada Rumora
Faculty of Pharmacy and Biochemistry, University of Zagreb,
Zagreb, Croatia

ABSTRACT

Fumonisin B_1 (FB_1) is a mycotoxin and a secondary product of *Fusarium* spp. molds. The predominant producer of FB_1 is *F. verticillioides* (formerly known as *F. moniliforme*). *Fusarium* spp. molds and FB_1 are common contaminants of maize worldwide. The toxicity of FB_1 was first observed in farm animals; in horses, exposure to FB_1-contaminated feed has been associated with equine leukoencephalo-malacia, and in pigs with pulmonary edema. For experimental animals, FB_1 is hepatotoxic, nephrotoxic and carcinogenic. In some regions of the world where maize is a staple food, increased incidences of human esophageal and liver cancers as well as increased frequency of neural tube defects (NTDs) have been observed and related to FB_1 exposure. NTDs are congenital malformations that occur as a result of failure of neural tube closure during early embryogenesis; in humans, the most common NTDs are anencephaly and spina bifida. Several lines of evidence point to an association between FB_1 and development of NTDs.

---

* E-mail: adomijan@pharma.hr.

It has been well-established that FB_1, due to its structural similarities with sphingoid bases sphinganine and sphingosine, alters sphingolipid metabolism. That in turn affects cell signal transduction pathways and folate receptor functioning. Therefore, FB_1-provoked impairment of sphingolipid metabolism has been suggested as a possible mechanism by which FB_1 induces the development of NTDs. Experimental studies have confirmed the embryotoxicty and teratogenicity of FB_1. Epidemiological studies demonstrated that maternal exposure to FB_1 (through consumption of FB_1-contaminated maize-based food) could be implicated in higher NTD occurrences. In this review, data associating exposure to FB_1 with the development of NTDs are presented.

Keywords: Fumonisin B_1, mycotoxin, neural tube defects, sphinganine, sphingosine, sphingolipids, folic acid

INTRODUCTION

Fumonisins are a group of mycotoxins and represent secondary metabolites of *Fusarium* spp. molds. They are produced primarily by *F. verticillioides* (formerly known as *F. moniliforme*) (Soriano and Dragacci, 2004; Voss et al., 2007). Thus far, more than 15 types of fumonisins have been identified and characterized, but fumonisin B_1 (FB_1) is the most abundant and toxicologically the most significant one (IPCS, 2001).

Fusarium spp. molds and fumonisins are found as contaminants in various commodities throughout the world. The most significant is contamination of maize with FB_1 (Soriano and Dragacci, 2004). In maize-based food (cornflakes or tortillas), lower levels of FB_1 are detected. This occurs because food processing (*e.g.* use of high temperature) degrades FB_1. Also, treatment of maize with alkaline solution during preparation of tortillas (nixtamalization) transforms FB_1 to hydrolyzed FB_1 (HFB_1, also known as aminopentol, AP1) (Soriano and Dragacci, 2004; Collins et al., 2006).

The toxic effects of FB_1 were first observed in domestic animals. In horses, the ingestion of FB_1-contaminated feed induced a neurotoxic syndrome known as equine leukoencephalomalacia (ELEM) and in pigs pulmonary edema (PPE) (Wilson et al., 1990; Voss et al., 2007). For experimental animals, FB_1 is hepatotoxic, nephrotoxic and carcinogenic. When laboratory rodents were chronically fed with FB_1, a significant increase in kidney tumors in male F344 rats and liver tumors in male BD IX rats and female $B6C3F_1$ mice were observed (NTP, 2001; Voss et al., 2007). Based on the available toxicological evidence, the International Agency for Research on Cancer (IARC) classified FB$_1$ in group 2B as a possible carcinogen to humans (IARC, 2002). According to toxicokinetic studies, FB$_1$ is poorly absorbed from gastrointestinal tract, rapidly eliminated from plasma and not extensively accumulated in tissues (Voss et al., 2007; Domijan, 2012). FB$_1$ is not metabolized by cytochrome P450 oxidases and transferases and is eliminated by feces nearly unchanged. Gut microbial flora can metabolize FB$_1$ to its hydrolyzed form (Voss et al., 2007; Muller et al., 2012).

Toxicity of FB$_1$ is the consequence of disruption of sphingolipid metabolism by FB$_1$. Due to its chemical structure and similarities with sphingoid bases sphingosine (So) and sphinganine (Sa) (Figure 1), FB$_1$ inhibits ceramide synthase which is the key enzyme in sphingolipid metabolism (Wang et al., 1991; Dragan et al., 2001; Desai et al., 2002). As an immediate consequence of this inhibition, the reversible accumulation of free Sa, and to a lesser extent So, occurs, and *de novo* biosynthesis of ceramides and complex sphingolipids is decreased (Figure 2). Free Sa and So as well as Sa/So ratio are considered to be good biomarkers of FB$_1$ exposure, but only in experimental conditions. As humans are exposed to much lower FB$_1$ concentrations, and the change in sphingoid base level is reversible, Sa, So and Sa/So ratio are not reliable biomarkers for population studies (Voss et al., 2007; Domijan, 2012).

The effect of FB$_1$ on human health is still unclear. In some regions of the world where maize is a staple food, higher incidences of esophageal cancer, liver cancer and neural tube defects (NTDs) occur which may be related to FB$_1$ exposure. However, epidemiological studies aiming to associate FB$_1$ with the development of esophageal or liver cancers are limited and do not offer clear conclusions (Wild and Gong, 2010; Muller et al., 2012). In contrast, some epidemiological and experimental studies have shown an association between maternal exposure to FB$_1$ and development of NTDs. NTDs are the second most common type of birth defect after congenital heart defects that occur due to failure of closure of embryonic neural tube during the first weeks of development (between 21 and 28 days after conception in humans) (Detrait et al., 2005; Mitchell, 2005). In humans, the most common NTDs are anencephaly and spina bifida. Anencephaly results from the failure of fusion of the cranial neural tube while spina bifida (myelomeningocele) results from failure of fusion in the spinal region of the neural tube (Detrait et al., 2005). The prevalence of NTDs depends on ethnicity and geographical location (Mitchell, 2005). This review presents findings that associate exposure to FB$_1$ during pregnancy with the development of NTDs.

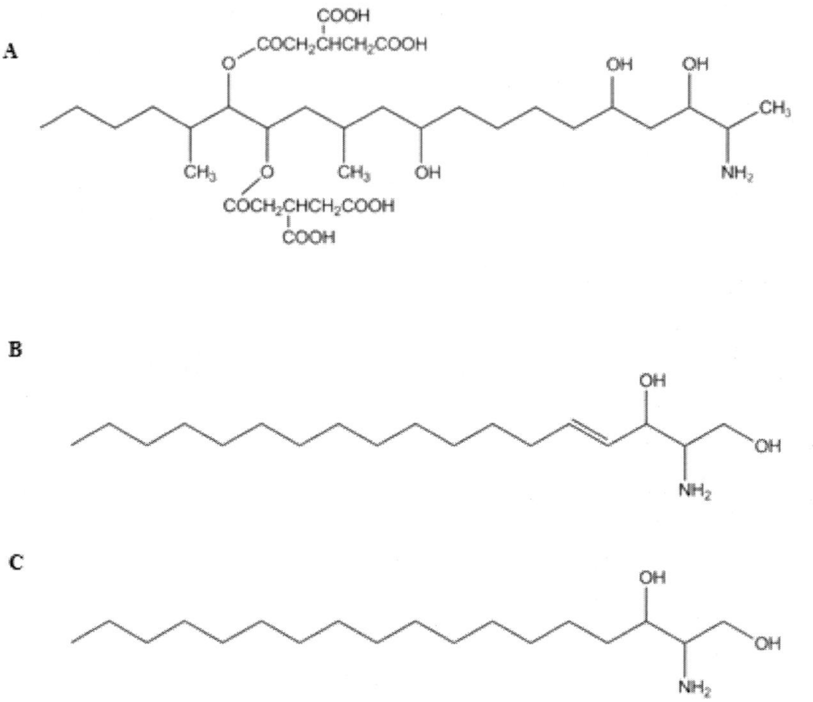

Figure 1. Structural formula of (A) fumonisin B_1, (B) sphingosine and (C) sphinganine.

MECHANISM OF ACTION

At the cellular level, several mechanisms of FB_1 toxicity have been established, such as altered sphingolipid metabolism, oxidative stress and disturbance in calcium signaling (Wang et al., 1991; Domijan and Abramov, 2011; Muller et al., 2012; Domijan, 2012). Since altered sphingolipid metabolism is observed in animal tissue or cells in culture immediately after treatment with FB_1, sphingolipid metabolism is considered the most significant mechanism that underlies FB_1 toxicity. Due to chemical structure similarities between FB_1 and sphingoid bases (Figure 1), FB_1 inhibits enzyme ceramide synthase, leading to disturbances in sphingolipid metabolism (Figure 2). Sphingolipids have several important functions within the cell. They are major structural components of cell membranes, but sphingolipids and their metabolites are also involved in signal transduction, thus leading to either cell death or survival.

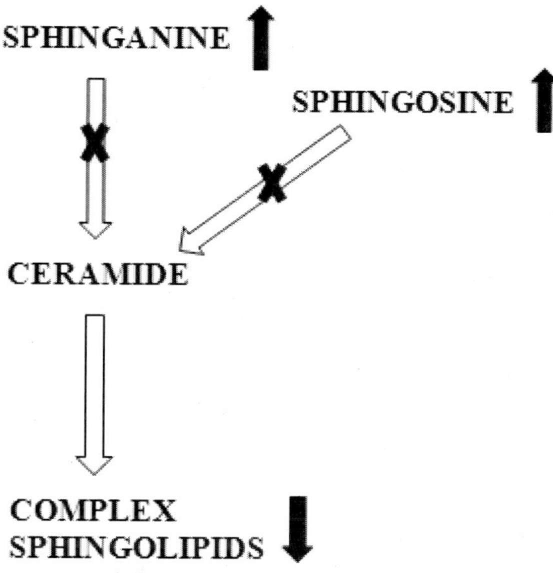

Adopted from Voss et al., 2007.

Figure 2. Simplified diagram of sphingolipid metabolism and sphingolipid turnover influenced by FB$_1$. FB$_1$-provoked inhibition of ceramide synthase activity (crossed white arrows) leads to down-regulation of *de novo* ceramide and complex sphingolipids synthesis. Black arrows indicate an increase or decrease in a metabolite's concentration.

In addition, sphingoid base phosphorylated metabolites Sa 1-phosphate and So 1-phosphate (that are increased after FB$_1$ exposure) regulate cellular processes such as cell survival, growth and differentiation through binding to So 1-phosphate receptors (S1P receptors). The functional S1P receptors are expressed in the embryonic brain in areas where active neurogenesis takes place (Hait et al., 2006). Mizugishi et al. (2005) showed that S1P signaling during embryonic development is critical for neurogenesis; *S1P$_1$* receptor-null mice showed severe defects in neurogenesis. The same study also demonstrated that *SphK*-null mice (sphingosin kinase catalyzes the phosphorylation of sphingosine to So 1-phosphate) exhibited a high level of NTDs (Mizugishi et al., 2005). In addition, in SWV and LM/Bc strains of mice it was found that both FB$_1$ and TY720 (an agonist of S1P receptor) induced NTDs, indicating that activation of S1P receptor signaling pathways could be involved in NTD development (Gelineau-van Waes et al., 2012). The authors concluded that elevated concentrations of sphingoid base 1-

phosphates, formed after FB_1 exposure, may perturb signaling cascades involved in embryonic morphogenesis by acting as ligands of S1P receptors.

It was shown that, by altering sphingolipid metabolism, FB_1 affects functions of folate receptor, thereby affecting the transport of folic acid. The high-affinity folate receptor (Folbp1 - murine; folate receptor α - human) is a glycosylphosphatidylinositol-anchored protein found in membrane microdomains rich in cholesterol and sphingolipids (Marasas et al., 2004; Gelineau-van Waes et al., 2005). In a study by Stevens and Tang (1997), FB_1 was demonstrated to inhibit folate uptake by folate receptor, and in turn compromise cellular processes dependent on this vitamin. In this study, FB_1 treatment of Caco-2 cells reduced the levels of sphingolipids, the rate of folate uptake and the total amount of folic acid binding capacity by folate receptor, indicating that sphingolipids play an important role in folate receptor function (Stevens and Tang, 1997). In a study on LM/Bc mice treated with FB_1 the distribution and abundance of folate receptors (Folbp1) in placenta and embryos were monitored by fluorescent immunolabeling (Gelineau-van Waes et al., 2005). FB_1 treatment dramatically reduced the number of folate receptors in the neuroepithelial cells and yolk sac membrane of embryos. Additionally, FB_1 treatment decreased the level of folate in embryonic tissues and placental tissue, which confirms an inhibition of folate uptake by FB_1. Since dose-dependent NTDs were observed after FB_1 treatment, a good association between FB_1 exposure, folate deficiency and increased risk for NTDs could be made. Therefore, it is not surprising that maternal periconceptional folic acid supplementation in prevention of NTDs is today well-documented and recommended.

EXPERIMENTAL STUDIES

The embryotoxicity of FB_1 was demonstrated in studies on rats and chick embryos (Flynn et al., 1996; Zacharias et al., 1996). FB_1 inhibited the growth and development of rat embryos (Flynn et al., 1996), while hemorrhages under the skin and in the liver were observed in FB_1-treated chick embryos (Zacharias et al., 1996).

Developmental toxicity of FB_1 was tested on mice, rats and rabbits (Reddy et al., 1996; Collins et al., 1998; LaBorde et al., 1997). In a study on CD1 mice treated with FB_1 (gestation days 7-15), offspring exhibited a dose-dependent increase in the incidence and severity of hydrocephalus of both the lateral and third ventricles. These results suggested that FB$_1$ may be a developmental toxicant.

However, maternal toxicity (hepatotoxicity) and alteration in Sa/So level in maternal, but not in fetal liver was also observed (Reddy et al., 1996). In a study on rats treated with FB$_1$ (gestation days 3–16), both maternal toxicity (inappetence, emaciation, lethargy, death) and fetal toxicity (decreased fetal body weight, decreased crown–rump length, increased the incidence of hydrocephalus and skeletal anomalies, increased the number of late deaths) were seen (Collins et al., 1998). Again, an increase of Sa/So was observed in maternal tissues, but not in fetal. In a study on New Zealand white rabbits treated with FB$_1$ (gestation day 3-19), a decrease in total fetal weight as well as in fetal liver and kidneys weight was observed. However, again Sa/So augmentation was only recorded in maternal, and not in fetal tissues (LaBorde et al., 1997). The authors concluded that FB$_1$ did not cross the placenta and was probably not teratogenic. These results can be attributed to different experimental settings (animal model, mode of application and rout of mycotoxin absorption, duration of treatment, and doses of FB$_1$ used).

On the other hand, Sadler et al. (2002) clearly demonstrated that FB$_1$ is embryotoxic and can produce NTDs. After long-term exposure (26 hours) of neurulating mouse embryos to FB$_1$, growth retardation and NTDs were observed. The incidence of NTDs was dose-dependent, and NTDs were cranial inasmuch that neural folds had failed to fuse in the mid line in one or more region of the forebrain, midbrain and hindbrain. Sa accumulation and an increase of Sa/So occurred in all of the embryos. Short-term exposure (2 hours) of neurulating mouse embryos to FB$_1$ caused growth inhibition and induced 66.7% of NTDs. Interestingly, co-treatment with folic acid in both exposure times (2 and 26 hours) reduced the incidence of NTDs, but had no effect on sphingoid base level meaning that FB$_1$ by disturbing embryonic sphingolipid metabolism interfered with folate utilization (Marasas et al., 2003). In a study on LM/Bc mice treated with FB$_1$ (on gestation day 7.5 and on gestation day 8.5, corresponding to early neurulation events in the mouse embryo), a dose-dependent increase in incidence of NTDs in fetuses was observed, from 5% with 5 mg FB$_1$/kg b.w. to 80% with 20 mg FB$_1$/kg b.w. (Gelineau-van Waes et al., 2005). Supplementation with folic acid only partially decreased the appearance of NTDs. These experiments confirmed that maternal exposure to FB$_1$ during early pregnancy, *i.e.* during early neurulating events in embryos, results in a higher NTD frequency.

Several studies tested the teratogenic potential of FB$_1$ derivative AP1 as well (Flynn et al. 1997; Collins et al., 2006; Voss et al., 2009). On cultured rat embryos, a higher doses of AP1 increased the incidence of abnormal embryos, suggesting that AP1 can induce NTDs. However, AP1 was 100-fold less toxic than FB_1 (Flynn et al., 1997). In a study on pregnant rats treated with AP1 (gestation days 3-16), reduced dam body weight was observed. Still, AP1 was not teratogenic, did not affect the sphingid base ratio, did not alter reproduction or development of fetuses and produced no dose-related histopathological effects in dams, indicating once again that it was less toxic than FB_1 (Collins et al., 2006). AP1 application was not able to produce NTDs in sensitive LM/Bc mouse model, and only a weak disturbance in sphinglipid metabolism was observed (Voss et al., 2009). Altogether, the results showed that AP1 toxicity was significantly decreased as compared to FB_1.

EPIDEMIOLOGICAL STUDIES

The first association between NTDs and maternal exposure to FB_1 was noted in the US in the 1990s. During 1990-1991, in a south Texas county bordering Mexico, a high level of anencephalic infants was born among Mexican-American women (27 per 10,000 live births compared to 15 per 10,000 live births during 1986-1989) (Hendricks, 1999). Just before these findings (in the fall of 1989), an outbreak of ELEM and PPE occurred in Texas and high levels of FB_1 in maize were detected. Therefore it is possible that a high incidence of NTDs along Texas-Mexico border area in 1990-1991 might be related to maternal exposure to FB_1 through FB_1-contaminated maize-based food (Hendricks, 1999; Missmer et al., 2006). Similarly, in other regions of the world where maize represents a staple food, higher incidences of NTDs have been recorded and associations to FB_1 exposure made (Marasas at al., 2004). For example in South Africa's rural regions Transkei and Limpopo, higher NTDs incidences (61 per 10,000 and 35 per 10,000, respectively) compared to urban regions such as Cape Town (12 per 10,000) or Pretoria (10 per 10,000) were observed (Ncayiyana, 1986; Venter et al., 1995; Delport et al., 1995). Additionally, in rural areas in the northern provinces of China, the prevalence rate of NTDs is among the highest in the world, about 60 per 10,000 live births (Moore et al. 1997; Marasas et al., 2004). A study conducted in Guatemala during 2000 observed a higher incidence of NTDs in the rural region Quetzaltenango (106 per 10,000 live births) compared to other parts of Guatemala (20 per 10,000) (Marasas et al., 2004; Nguyen et al., 2008; Manucci and von Quednow, 2014).

Only one case-control study designed in order to identify whether maternal exposure to FB$_1$ increases the risk of NTDs in offspring has been published so far (Missmer et al., 2006). In this study, performed from March 1995 to May 2000, Mexican-American women with NTD-affected pregnancies who had resided and given birth along the Texas-Mexican border were recruited. Age-matched Mexican-American women from the same region who delivered normal live births were selected as controls. An interview, a questionnaire on maize tortilla consumption before conception and during the first trimester, and blood collection for serum analysis of maternal Sa/So ratio (a possible biomarker for FB$_1$ exposure) as well as serum folate and B$_{12}$ levels were conducted postpartum. The recruited women had low socioeconomic status (only a few reported taking preconceptional folic acid supplementation: 4.4% control women and 6% women with NTDs); however control women were leaner and more educated than the case women. The results showed that after adjusting for BMI, serum B$_{12}$ and date of conception, moderate consumption of tortillas during the first trimester was associated with increased odds ratios for NTD occurrence. However, higher intakes of tortilla had no effect on NTD occurrence. Similarly, higher Sa/So ratio was associated with increased odds ratios for NTD development, but the highest Sa/So ratio was not. Therefore, the authors concluded that maternal FB$_1$ exposure increase the risk for NTDs up to a threshold level, at which point fetal death is more likely to occur.

CONCLUSION

NTDs are multifactorial disorders, probably involving genetic, nutritional and environmental factors. Several lines of evidence link FB$_1$ exposure to the development of NTDs. The mechanism of FB$_1$ action through altered sphingolipid metabolism, which in turn affects cellular signaling and alters the function of folate receptor, is suggested. Experimental studies demonstrated FB$_1$ embryotoxicity and teratogenicity. Moreover, an epidemiological study revealed an association between consumption of FB$_1$-contaminated maize-based food and development of NTDs. Therefore maternal exposure to FB$_1$ should be considered a potential risk factor for NTD occurrence, together with other well-established etiological factors for NTDs.

REFERENCES

Collins T. F. X., Sprando R. L., Black T. N., Shackelford M. E., Laborde J. B., Hansen D. K., Eppley R. M., Trucksess M. W., Howard P. C., Bryan M. A., Ruggles D. I., Olejnik N., Rorie J. I. Effects of fumonisin B_1 in pregnant rats. Part 2. *Food Chem. Toxicol.*, 1998; 36: 673–685.

Collins T. F. X., Sprando R. L., Black T. N., Olejnik N., Eppley R. M., Shackelford M. E., Howard P. C., Rorie J. I., Bryant M., Ruggles D. I. Effects of aminopentol on in utero development in rats. *Food Chem. Toxicol.*, 2006; 44: 161–169.

Delport S. D., Christianson A. L., van den Berg H. J., Wolmarans L., Gericke G. S. Congenital anomalies in black South African liveborn neonates at an urban academic hospital. *S. Afr. Med. J.*, 1995; 85: 11–15.

Desai K., Cameron Sullards M., Allegood J., Wang E., Schmelz E. M., Hartl M., Humpf H.-U., Liotta D. C., Peng Q., Merrill A. H., Jr. Fumonisins and fumonisin analogs as inhibitors of ceramide synthase and inducers of apoptosis. *Biochim. Biophys. Acta*, 2002; 1585: 188–192.

Detrait E. R., George T. M., Etchevers H. C., Gilbert J. R., Vekemans M., Speer M. C. Human neural tube defects: Developmental biology, epidemiology and genetics. *Neurotoxiciol. Teratol.*, 2005; 27: 515–524.

Domijan A.-M., Abramov A. Y. Fumonisin B_1 inhibits mitochondrial respiration and deregulates calcium homeostasis – implication to mechanism of cell toxicity. *Int. J. Biochem. Cell Biol.*, 2011; 43: 897-904.

Domijan A.-M. Fumonisin B1: a neurotoxic mycotoxin. *Arh. Hig. Rada. Toxicol.*, 2012; 63: 531-544.

Dragan Y. P., Bidlack W. R., Cohen S. M., Goldsworthy T. L., Hard G. C., Howard P. C., Riley R. T., Voss K. A. Implications of apoptosis for toxicity, carcinogenicity and risk assessment: fumonisin B_1 as an example. *Toxicol. Sci.*, 2001; 61: 6–17.

Flynn T. J., Pritchard D., Bradlaw J., Eppley R., Page S. *In vivo* embryotoxicity of fumonisin B1 evaluated with cultured postimplantation staged rat embryos. *In Vitro Toxicol.*, 1996; 9: 271–279.

Flynn T. J., Stack M. E., Troy A. L., Chirtel S. J. Assessment of the embryotoxic potential of the total hydrolysis product of fumonisin B_1 using cultured organogenesis-staged rat embryos. *Food Chem. Toxicol.*, 1997; 35: 1135–1141.

Gelineau-van Waes J., Starr L., Maddox J. R., Aleman F., Voss K. A., Wilberding J., Riley R. T. Maternal fumonisin exposure and risk for neural tube defects: mechanisms in an in vivo mouse model. *Birth Defects Res. (Part A): Clin. Mol. Teratol.*, 2005; 73: 487–497.

Gelineau-van Waes J., Rainey M. A., Maddox J. R., Voss K. A., Sachs A. J., Gardner N. M., Wilberding J. D., Riley R. T. Increased sphingoid base-1-phosphates and failure of neural tube closure after exposure to fumonisin or FTY720. *Birth Defects Res. A. Clin. Mol. Teratol.*, 2012; 94: 790–803.

Hait N. C., Oskeritzian C. A., Paugh S. W., Milstien S., Spiegel S. Sphingosine kinases, sphingosine 1- phosphate, apoptosis and diseases. *Biochem. Biophy. Acta*, 2006; 1758; 2016–2026.

Hendricks K. Fumonisins and neural tube defects in South Texas. *Epidemiology*, 1999; 10: 198–200.

International Agency for Research on Cancer, IARC. Some traditional herbal medicines, some mycotoxins, naphthalene and styrene. IARC monographs on the evaluation of carcinogenic risks to humans, 2002: Vol. 82. Lyon, International Agency for Research on Cancer.

International Programme on Chemical Safety, IPCS. Safety Evaluation of Certain Mycotoxins in Food. WHO Food Add. Series. 2001: Vol. 47. Geneva, WHO.

LaBorde J. B., Terry K. K., Howard P. C., Chen J. J., Collins T. F. X., Shackelford M. E., Hansen D. K. Lack of embryotoxicity of fumonisin B1 in New Zealand white rabbits. *Fundam. Appl. Toxicol.*, 1997; 40: 120–128.

Manucci G., von Quednow E. How I do it: neural tube defects in Guatemala – myelomeningocele spina bifida unit. *Surg. Neurol. Int.*, 2014; 5: S13–22.

Marasas W. F. O., Riley R. T., Hendricks K. A., Stevens V. L., Sadler T. W., Gelineau-van Waes J., Missmer S. A., Cabrera J., Torres O., Gelderblom W. C. A., Allegood J., Martinez C., Maddox J., Miller J. D., Starr L., Cameron Sullards M., Roman A. V., Voss K. A., Wang E., Merrill A. H., Jr. Fumonisins distrupt sphingolipid metabolism, folate transport and neural tube development in embryo culture and in vivo: a potential risk factor for human neural tube defect among populations consuming fumonisin-contaminated maize. *J. Nutrit.*, 2004; 134: 711–716.

Missmer S. A., Suarez L., Felkner M., Wang E., Merrill A. H., Jr., Rotman K. J., Hendricks K. A. Exposure to fumonisins and the occurrence of neural tube defects along Texas Mexico border. *Environmen. Health Persp.*, 2006; 114: 237–241.

Mizugishi K., Yamashita T., Olivera A., Miller G. F., Spiegel S., Proia R. L. Essential role for sphingosine kinase in neural and vascular development. *Mol. Cell Biol.*, 2005; 25: 11113–11121.

Mitchell L. E. Epidemiology of neural tube defects. *Am. J. Med. Genet. C. Semin. Med. Genet.,* 2005; 135C: 88–94.

Moore C. A., Li S., Li Z., Hong S. X., Gu H. Q., Berry R. J., Mulinare J., Erickson J. D. Elevated rates of severe neural tube defects in a high-prevalence area in northern China. *Am. J. Med. Genet.,* 1997; 73: 113–118.

Müller S., Dekant W., Mally A. Fumonisin B1 and the kidney: Modes of action for renal tumor formation by fumonisin B_1 in rodents. *Food Chem. Toxicol.,* 2012; 50: 3833–3846.

National Toxicology Program, NTP Technical Report on the Toxicology and Carcinogenesis Studies of fumonisin B_1 (CAS No. 116355-83-0) in F344/N rats and B6C3F1 mice. Research Triangle Park, 2001: NC: National Toxicology Program. http://ntp.niehs.nih.gov/ntp/htdocs/LT_rpts/tr496.pdf

Ncayiyana D. J. Neural tube defects among rural blacks in a Transkei district. A preliminary report and analysis. *S. Afr. Med. J.,* 1986; 69: 618–620.

Nguyen P., Grajeda R., Melgar P., Marcinkevage J., Flores R., Martorell R. Weekly may be as efficacious as daily folic acid supplementation in improving folate status and lowering serum homocysteine concentrations in Guatemalan women. *J. Nutr.,* 2008; 138: 1491–1498.

Reddy R. V., Johnson G., Rottinghaus G. E., Casteel S. W., Reddy C. S. Developmental effects of fumonisin B_1 in mice. *Mycopathologia,* 1996; 134: 161-166.

Sadler T. W., Merrill A. H., Stevens V. L., Cameron Sullards M., Wang E., Wang P. Prevention of fumonisin B1-induced neural tube defects by folic acid. *Teratology,* 2002; 66: 169–176.

Stevens V. L., Tang J. Fumonisin B1-induced sphingolipid depletion inhibits vitamin uptake via the glycosylphosphatidylinositol-anchored folate receptor. *J. Biol. Chem.,* 1997; 272: 18020–18025.

Soriano J. M., Dragacci S. Occurence of fumonisins in food. *Food Res. Inter.,* 2004; 37: 985–1000.

Venter P. A., Christianson A. L., Hutamo C. M., Makhura M. P., Gericke G. S. Congenital anomalies in rural black South African neonates--a silent epidemic? *S. Afr. Med. J.,* 1995; 85: 15–20.

Voss K. A., Smith G. W., Haschek W. M. Fumonisins: Toxicokinetics, mechanism of action and toxicity. *Animal Feed Sci. Technol.,* 2007; 137: 299–325.

Voss K. A., Riley R. T., Snook M. E., Gelineau-von Waes J. Reproductive and sphingolipid metabolic effects of fumonisin B1 and its alkaline hydrolysis product in LM/Bc mice: hydrolyzed fumonisin B1 did not cause neural tube defects. *Toxicol. Sci.,* 2009; 112: 459–467.

Wang E., Norred W. P., Bacon C. W., Riley R. T., Merrill A. H., Jr. Inhibition of sphingolipid biosynthesis by fumonisins. *J. Biol. Chem.,* 1991; 266: 14486–14490.

Wild C. P., Gong Y. Y. Mycotoxins and human disease: a largely ignored global health issue. *Carcinogenesis,* 2010; 31: 71–82.

Wilson T. M., Ross P. F., Rice L. G., Osweiler G. D., Nelson H. A., Owens D. L., Plattner R. D., Reggiardo C., Noon T. H., Pickrell J. W. Fumonisin B$_1$ levels associated with an epizootic of equine leukoencephalomalacia. *J. Vet. Diagn. Invest.,* 1990; 2: 213–216.

Zacharias C., van Echten-Deckert G., Wang E., Merrill A. H., Sandhoff K. The effect of fumonisin B$_1$ on developing chick embryos: correlation between de novo sphingolipid biosynthesis and gross morphological changes. *Glycocony J.,* 1996; 13: 167–175.

In: Fumonisins
Editor: Craig M. Evans

ISBN: 978-1-63482-789-8
© 2015 Nova Science Publishers, Inc.

Chapter 3

STATE-OF-THE-ART IN THE ANALYSIS OF FUMONISINS BY LIQUID CHROMATOGRAPHY-MASS SPECTROMETRY

Paolo Lucci[1], Rocío Castro-Ríos[2] and Oscar Núñez[3,4,*]
[1]Department of Nutrition and Biochemistry, Faculty of Sciences,
Pontificia Universidad Javeriana, Bogotá, Colombia
[2]Department of Analytical Chemistry, Faculty of Medicine, Universidad
Autónoma de Nuevo León. Av. Madero s/n, Mitras Centro Monterrey,
Nuevo León, Mexico
[3]Department of Analytical Chemistry,
University of Barcelona. Barcelona, Spain
[4]Serra Húnter Fellow, Generalitat de Catalunya, Spain

ABSTRACT

Fumonisins are mycotoxins produced by various *Fusarium* spp. Fungi (mainly *Fusarium verticillioides* and *Fusarium proliferatum*) commonly found in maize. They were first discovered in the late 80s by Gelderblon et al., who studied the incidence rate of esophageal cancer in the South African population. More than ten different fumonisins have been isolated and characterized. Of these, fumonisins B_1, B_2 and B_3 (FB_1, FB_2 and FB_3, respectively) are the major ones naturally occurring in

---

[*] E-mail address: oscar.nunez@ub.edu.

maize or maize-based products. Fumonisins are toxic toward many different animals. Known animal diseases caused by these mycotoxins are hepatocarcinogenic effects of rats and mice, equine leukoencephalomalacia in horses, and pulmonary edema in pigs. However, their toxicity toward humans is still unclear. There is some evidence that fumonisins might be involved in the formation of esophageal cancer and might have a negative effect on neural tube development in embryos. The International Agency for Research on Cancer classified FB_1 as a Group 2B carcinogen (possibly carcinogenic to humans). For these reasons, maximum levels for fumonisin contamination have been decreed by many countries. For instance, in the European Union, the maximum permitted levels for fumonisins (sum of FB_1 and FB_2) in maize and its derived products are 0.2 to 4 mg/Kg. Therefore, the development of analytical methodologies for the routine control of fumonisins in food is necessary.

Various methodologies have been employed for the analysis of fumonisins. Although fumonisins molecules neither selectively absorb UV light nor exhibit any fluorescence, since they lack chromophores in their molecular structure, liquid chromatography (LC) with fluorescence detection have been employed after fumonisin derivatization into compounds that exhibit fluorescence. But nowadays, liquid chromatography-mass spectrometry (LC-MS(/MS)) and liquid chromatography-high resolution mass spectrometry (LC-HRMS) are the techniques of choice for the analysis and characterization of fumonisins in foodstuffs. However, it should be pointed out that regardless of the employed detection system, sample preparation procedures (especially extraction and clean-up procedures) is maybe the most important issue in the determination of fumonisins. Various extraction techniques have been reported for the clean-up and extraction of fumonisins including simple solid-liquid extraction (SLE), solid-phase extraction (SPE), immunoaffinity procedures, the use of molecularly imprinted polymers (MIPs), and even QuEChERS methods.

This chapter will review the state-of-the-art of liquid chromatography-mass spectrometry techniques for the analysis and characterization of fumonisins in food-based products and other matrices. Because of their importance in fumonisin analysis, commonly used as well as novel sample treatment procedures will be addressed. LC-MS chromatographic conditions, ionization sources, and MS and HRMS analyzers frequently used, as well as strategies for the structural characterization and the qualitative and quantitative analysis of fumonisins will be discussed by means of relevant applications. Coverage of all kind of applications is beyond the scope of the present contribution, so we will focus on the most relevant applications published in the last years.

INTRODUCTION

Mycotoxins are secondary metabolites produced by different types of moulds (*Aspergillus, Penicillium* or *Fusarium* species) [1, 2]. They are produced by fungal infection of agricultural crops in the field or during harvest, drying or subsequent storage. In general, mycotoxins are very stable compounds that cannot be readily destroyed by heating or during food processing, although there can be reductions in their levels, for example, during milling of grain. Today, more than 400 mycotoxins are known, being aflatoxins, ochratoxins, patulin, zearalenone, fumonisins and trichothecenes the most important ones for public health [3].

Fumonisins (FBs), a family of food-borne mycotoxins with a wide spectrum of toxicological activities, were first isolated in the 1988s from cultures of *Fusarium verticillioides* by Gelderblom et al., [4, 5] who studied the incidence rate of esophageal cancer in the South African population. The 28 fumonisins analogues that have been characterized since 1988 can be separated into four main groups, identified as A, B, C and P series, of which the toxicologically most important fumonisins are the FB analogues (see structures in Figure 1). As can be seen, they consist of a C_{20} backbone structure which bears hydroxyl groups at different positions. Among them, Fumonisin B_1, B_2 and B_3 (FB_1, FB_2 and FB_3, respectively) are the most abundant occurring fumonisins in nature. They can be found worldwide as contaminants of many grains and plant-derived foods [6], and FB_1 is usually found at high levels most frequently in maize and in maize-based foodstuffs and feedstuffs. In fact, it is estimated that more than half of maize- and maize-based products produced around the world are contaminated with fumonisins [7-9]. However, fumonisins have also been found in other plants such as white beans, wheat, barley, soy beans, figs, black tea, and even in medicinal herbs [10-13].

Fumonisins are toxic toward many different animals. Known animal diseases caused by these mycotoxins are hepatocarcinogenic effects of rats and mice, equine leukoencephalomalacia in horses, and pulmonary edema in pigs. However, their toxicity toward humans is still unclear. There is some evidence that fumonisins might be involved in the formation of esophageal cancer and might have a negative effect on neural tube development in embryos. The International Agency for Research on Cancer classified FB_1 as a Group 2B carcinogen (possibly carcinogenic to humans).

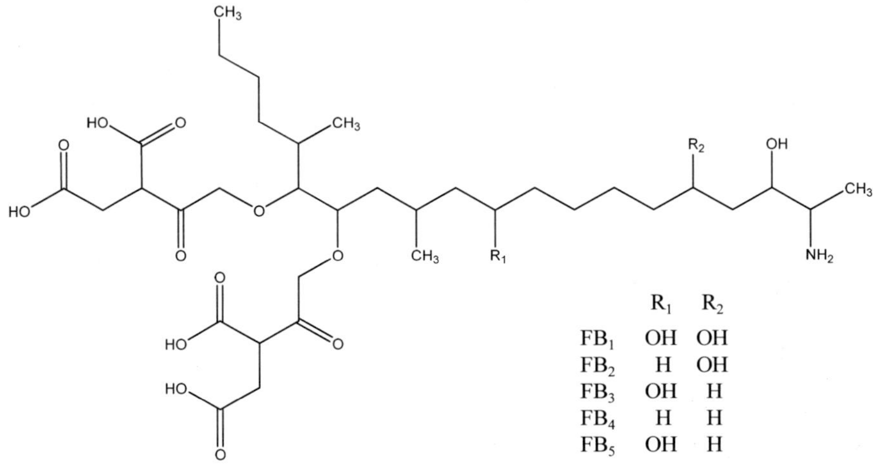

	R_1	R_2
FB_1	OH	OH
FB_2	H	OH
FB_3	OH	H
FB_4	H	H
FB_5	OH	H

Figure 1. Structure of the main fumonisin B series.

However, corn contamination strongly depends on climatic conditions in the region where corn was grown. In countries with hot and wet climate maize is a basic food product crucial for alimentation of the population. Commonly observed corn contamination and high consumption rates may combine in such countries into a high health-related risk for population exposed to fumonisins from dietary intake. In contrast, the majority of corn grown in developed countries is allocated for fodder purposes; therefore, the risk to human health related to dietary intake of fumonisins is significantly low.

Nevertheless, maximum levels for fumonisin contamination have been decreed by many countries. For instance, in the European Union, the maximum permitted levels for fumonisins (sum of FB_1 and FB_2) in maize and its derived products are 0.2 to 4 mg/Kg [14]. US guidelines for total FB_1, FB_2 and FB_3 are 2 or 4 mg/Kg in maize products used for human foods and 3 mg/Kg for popcorn grain [15]. Thus, in order to guarantee food quality and safety, reliable analytical methodologies for the analysis of fumonisins are necessary.

To determine fumonisins in food and feed samples, many liquid chromatography (LC) methods with fluorescence detection (LC-FD) [16] have been proposed. However, all of the LC-FD methods require a derivatization step, and false results occur frequently because of retention time shifts or interfering peaks. Therefore, liquid chromatography coupled to mass spectrometry (LC-MS) has attracted greater attention for the analysis of fumonisins and mycotoxins due to its capacity for reliable analysis and small sample requirements. In this context, high resolution mass spectrometry (HRMS) is also playing an important role to help in the analysis and characterization of fumonisins in foodstuffs [17-20].

SAMPLE PREPARATION PROCEDURES

Despite the significant advances in chromatographic separations and MS techniques, sample treatment is still one of the most significant parts of the entire analytical process. Considerable efforts have therefore been made in the past years to simplify this step and to develop fast, accurate and precise methodologies able to allow the analytical determination of target mycotoxins without compromising the integrity of the extraction process.

Taking into consideration that fumonisins are usually detected at very low levels (ng/g) and that food and foodstuffs are very complex matrices, the choice of the proper sample treatment is a fundamental prerequisite for the obtainment of adequate clean-up and concentration prior to instrumental analysis. According to the literature, a multiplicity of sample preparation techniques has been used over the last few years for the extraction and purification of fumonisins in food matrices. However, regarding the extraction procedure, liquid extraction (LE) is probably the most popular technique, and because of the polar nature of fumonisins, a mixture of acidified solvents (e.g., methanol-water [21, 22], acetonitrile-water [23, 24], and methanol-acetonitrile-water [25]) are usually employed, obviously depending on the composition of the food matrix (Table 1). Other alternatives to conventional LE, such as pressurized liquid extraction (PLE) [26], and supercritical fluid extraction (SFE) [27], have also been reported with the aim of overcoming some drawbacks of traditional approaches (i.e., time and solvent consumption). For instance, D'Arco et al., [26] have recently described a method based on PLE for the extraction of fumonisins B_1, B_2 and B_3 in corn-based baby food. The method, which was optimized by the authors for several extraction parameters able to affect PLE efficiency (i.e., pressure, solvent extraction, number of cycles and dispersant/clean-up agents), showed satisfactory recoveries ranging from 68 to 83% at fortification levels of 200 µg/kg with relative standard deviation (RSD) values from 4 to 12%.

But regardless of the technique used for the extraction, a clean-up procedure is usually required prior to chromatographic analysis in order to reduce the amount of co-extracted compounds, thus minimizing the matrix effects. Sorbent-based techniques are the most commonly adopted analytical approach for both extraction and clean-up purposes and several authors have reported the use of octadecyl- (C18) bonded silicas [23, 28, 29], hydrophilic–lipophilic balanced polymeric sorbents (Oasis-HLB®) [30], as well as anion exchangers SPE materials [31]. For instance, Seefelder et al., [32] investigated the use of SAX SPE column for the analysis of fumonisin B_1 in *fusarium proliferatum*-infected asparagus spears and garlic bulbs using a mixture of methanol/0.1 M HCl for extraction and the strong anion exchanger for samples clean-up. The method, which also involved the use of isotopically labeled FB_1-d_6 as internal standard, provided good extraction efficiencies with recoveries ranging from 96 to 104% for garlic samples and from 79 to 92% for asparagus samples. On the other hand, Oasis-HLB® cartridges have been recently successfully used for the analysis of fumonisins B_1, B_2 and their hydrolyzed metabolites in a complex matrix such as pig liver [30]. However, in this study, the author reported that the solvent evaporation required for concentrating the analytes before LC-MS/MS analysis led to a great variability in recovery, with unreproducible values ranging from 10 to 80%. To solve this problem the methanol eluate was just dried up to about 200 µL (and not completely dried) under a stream of nitrogen at 40°C and then reconstituted with mobile phase in 1 mL volumetric flask. Under these conditions, satisfactory recoveries >83% were obtained in all the experiments conducted.

A number of new sorbents have also been developed during the last decade in order to obtain higher SPE selectivity. As results, sophisticated antibodies and intelligent polymer such as molecularly imprinted polymers (MIP$_S$) are now available in the market for many applications, including fumonisins [25, 33]. Immunoaffinity column (IAC) has been widely reported in the literature to allow selective extraction, concentration and clean-up of fumonisins from complex matrices thanks to its retention mechanism based on antigen-antibody interactions. IACs have been effectively employed for the determination of fumonisins in corn-based product [34], cornflakes [21, 22], bovine milk [35], urine [36], and spices and aromatic medicinal herbs [37]. The use of MIP$_S$ has also been reported [25, 33]. For instance, AFFINIMIP FumoZON MIP SPE cartridges have recently been used for the determination of the B_1, B_2, and B_3 fumonisins in 49 cereal samples (42 maize-based and seven wheat-based products) showing high specificity and selectivity together with satisfactory extraction efficiency [25]. In fact, the accuracy of the method was found to range from 95 to 100% for the all studied fumonisins with an average RSD values lower than 17% in all cases. Interestingly, the developed MIP SPE method showed also almost equal selectivity against B_1, B_2, and B_3 forms.

Table 1. Selection of LC-MS(/MS) methods for the analysis of fumonisins

Compounds	Samples	Sample treatment	Chromatographic conditions	Mass spectrometry conditions	Analysis time	LOQs	Ref.
Fumonisin B_1 and N-(Carboxymethyl)-fumonisin B_1	Corn products	Extraction with methanol:acetonitrile:water (25:25:50 *v/v/v*) Clean-up with C18 cartridges	Waters Symmetry C18 column (150x2.1mm, 5 µm) *Gradient elution:* A) Water (0.05% trifluoroacetic acid) B) Acetonitrile (0.05% trifluoroacetic acid) 0.20 mL/min	Triple quadrupole mass analyzer (+) ESI SRM acquisition mode	10 min	10 µg/kg (LOD)	[28]
Fumonisins B_1	Asparagus spears and garlic bulbs	Extraction with methanol (0.1 M HCl) Clean-up with SAX Isolute SPE cartridges	Lichrospher 60-RP select B column (100x2.0 mm, 5 µm) for Asparagus Waters Symmetry C18 column (150x2.1 mm, 5 µm) for Garlic *Gradient elution:* A) Water (0.05% trifluoroacetic acid) B) Methanol (0.05% trifluoroacetic acid) for Asparagus B) Acetonitrile (0.05% trifluoroacetic acid) for Garlic 0.20 mL/min	Single quadrupole mass analyzer (+) ESI SIM acquisition mode	5 min (Asparagus) 7.5 min (Garlic)	26-36 µg/kg (LOD)	[32]
Fumonisins B_1, B_2 and hydrolysed FB_1	Microbial culture media	-	Agilent Eclipse XDB-C8 column (150x4.6 mm, 5 µm) *Gradient elution:* A) water (5 mM ammonium formate, pH 3) B) Acetonitrile 0.6 mL/min	Single quadrupole mass analyzer (+) ESI SIM acquisition mode	18 min	21-53 µg/L (LOD)	[45]

Table 1. (Continued)

Compounds	Samples	Sample treatment	Chromatographic conditions	Mass spectrometry conditions	Analysis time	LOQs	Ref.
Fumonisins B_1, B_2 and B_3	Cornflakes	Extraction with methanol:water (70:30 *v/v*) Clean-up with FumoniTest™ immunoaffinity column	Alltima C18 column (150x3.2 mm, 5 μm) *Isocratic elution:* Acetonitrile:water (60:40 *v/v*) with 0.3% formic acid 0.30 mL/min	Triple quadrupole mass analyzer (+) ESI SRM acquisition mode	8 min	15-40 μg/kg	[21, 22]
Fumonisins B_1 and B_2	Maize and maize-based products	Extraction with acetonitrile:water (75:25 *v/v*) with 50 mM formic acid Defatting with C18 phase. Clean-up with Carbograph-4 cartridges	Alltima C18 column (250x3.2 mm, 5 μm) *Gradient elution:* A) Water (25 mM formic acid) B) Methanol (25 mM formic acid) 0.20 mL/min	Quadrupole-Linear ion trap (+) ESI SRM acquisition mode	13 min	10-20 μg/kg (LOD)	[23]
Fumonisins B_1, B_2, B_3 and biomarkers of disrupted sphingolipid metabolism	Tissues of maize seedlings	Extraction with acetonitrile:water (50:50 *v/v*) with 5% formic acid	Metachem Inertsil ODS-3 column (150x3 mm, 5 μm) *Gradient elution:* A) Water:methanol:formic acid (97:2:1 *v/v/v*) B) Methanol:water:formic acid (97:2:1 *v/v/v*) 0.20 mL/min	Linear ion trap (+) ESI SRM acquisition mode	20 min	10 μg/kg (LOD)	[24]
Fumonisins B_1, B_2 and B_3	Corn-based baby food	Pressurized liquid extraction (PLE) Elution solvent: methanol	Luna C18 column (150x4.6 mm, 5 μm) *Gradient elution:* A) Water (0.5% formic acid) B) Methanol (0.5% formic acid) 0.30 mL/min	Triple quadrupole mass analyzer (+) ESI SRM acquisition mode	6 min	2-5 μg/kg	[26]

Compounds	Samples	Sample treatment	Chromatographic conditions	Mass spectrometry conditions	Analysis time	LOQs	Ref.
Fumonisins B_1 and B_2	Corn-based products	Extraction with methanol:water (80:20 *v/v*) Clean-up with FumoniTest™ immunoaffinity column	Luna C18 column (150x4.6 mm, 5 μm) *Gradient elution:* A) Water (0.5% formic acid) B) Methanol (0.5% formic acid) 0.50 mL/min	Triple quadrupole mass analyzer (+) ESI SRM acquisition mode	14 min	35 μg/kg	[34]
Fumonisins B_1, B_2 and B_3	Bell pepper, rice and corn flakes	Molecularly imprinted SPE cartridges	Alltima C18 column (150x3.2 mm, 5 μm) *Isocratic elution:* Water:Acetonitrile (60:40 *v/v*) with 0.3% formic acid 0.30 mL/min	Triple quadrupole mass analyzer (+) ESI SRM acquisition mode	-	9-44 μg/kg	[33]
Fumonisin B_1	Bovine milk	Vicam FumoniTest™ Immunoaffinity column	XTerra MS C18 column (150x2.15 mm, 5 μm) *Isocratic elution:* 75% of Water:acetonitrile (90:10 *v/v*) with 0.3% formic acid and 25% of acetonitrile (0.3% formic acid) 0.30 mL/min	Triple quadrupole mass analyzer (+) ESI SRM acquisition mode	4 min	0.1 μg/kg	[35]
Fumonisin B_1 and B_2	Food and feed	Extraction with acetonitrile:wáter (50:50 *v/v*) Clean-up with C18 SPE cartridge	Inertsil ODS3 column (50x2.1 mm, 3 μm) *Gradient elution:* A) Water (0.2% formic acid) B) Methanol (0.2% formic acid) 0.20 mL/min	Triple quadrupole mass analyzer (+) ESI SRM acquisition mode	6min	0.01-0.04 μg/kg (LOD)	[29]

Table 1. (Continued)

Compounds	Samples	Sample treatment	Chromatographic conditions	Mass spectrometry conditions	Analysis time	LOQs	Ref.
Fumonisin B_1 and B_2	Urine	Clean-up with FumoniTest™ immunoaffinity column	Luna C18 column (150x4.6 mm, 5 μm) *Gradient elution:* A) Water (0.5% formic acid) B) Methanol (0.5% formic acid) 0.50 mL/min	Triple quadrupole mass analyzer (+) ESI SRM acquisition mode	15 min	10 μg/L	[36]
Fumonisins B_1, B_2 and B_3	Maize	Extraction with acetonitrile:water (1:1 *v/v*) Clean-up with MultiSep 211 FUM SPE cartridges	UPLC BEH C18 column (100x2.1 mm, 1.7 μm) *Gradient elution:* A) Water (0.1% formic acid) B) Acetonitrile:methanol (1:1 *v/v*) 0.35 mL/min	Triple quadrupole mass analyzer (+) ESI SRM acquisition mode	3 min	0.4-1.65 μg/kg	[46]
Fumonisin B_1 and B_2, and the hydrolized metabolites	Pig liver	Extraction with methanol:water (80:20 *v/v*) Defatting with hexane Clean-up with Oasis HLB SPE cartridges	Waters XTerra MS C18 coumn (150x2.15 mm, 5 μm) *Gradient elution:* A) Acetonitrile:water (90:10 *v/v*) with 0.3% formic acid B) Acetonitrile (0.3% formic acid) 0.30 mL/min	Triple quadrupole mass analyzer (+) ESI SRM acquisition mode	12 min	10 μg/kg	[30]
Fumonisins B_1 and B_2	Corn	Ultrasonic extraction Methanol:water (3:1 *v/v*)	Zorbax Eclipse XDB-C18 column (150x2.1 mm, 3.5 μm) *Isocratic elution:* Methanol:water:formic acid (75:25:0.2 *v/v/v*) 0.20 mL/min	Triple quadrupole mass analyzer (+) ESI SRM acquisition mode	3 min	8.3-11.7 μg/kg	[47]

Compounds	Samples	Sample treatment	Chromatographic conditions	Mass spectrometry conditions	Analysis time	LOQs	Ref.
Fumonisins B_1 and B_2	Spices and aromatic and medicinal herbs	Extraction with methanol:water (80:20 *v/v*) Clean-up with FumoniTest™ immunoaffinity column	Gemini C18 column (20x2.0 mm, 3 µm) *Gradient elution:* A) Water (0.1% formic acid) B) Acetonitrile (0.1% formic acid) 0.50 mL/min	Quadrupole-ion trap (QTrap) mass analyzer (+) ESI SRM acquisition mode	2 min	100 µg/kg	[37]
Fumonisins B_1, B_2 and B_3	Cerereal products	Extraction with ACN/methanol/water (25:25:50 *v/v/v*) SPE with AFFINIMIP FumoZon MIP cartridges	Kinetex PFP column (100x2.1 mm, 2.6 µm) *Gradient elution:* A) Methanol:water:acetic acid (20:79.9:0.1 *v/v/v*) B) Methanol:water:acetic acid (79:19.9:0.1 *v/v/v*) 0.15 mL/min	Ion trap mass analyzer (+) ESI SRM acquisition mode	26 min	25 µg/kg	[25]
Fumonisin B_1	Maize	Extraction with acetonitrile:water (70:30 *v/v*) with 1% formic acid	Hyperclone C8 BDS column (150x2.0 mm, 3 µm) *Gradient elution:* A) Water (1% formic acid) B) Acetonitrile (1% formic acid) 0.30 mL/min	Triple quadrupole mass analyzer (+) ESI SRM acquisition mode	11 min	188 µg/kg	[48]

Table 2. Selection of LC-MS(/MS) multi-residue methods for the analysis of mycotoxins (including fumonisins)

Compounds	Samples	Sample treatment	Chromatographic conditions	Mass spectrometry conditions	Analysis time	LOQs[a]	Ref.
Fumonisin B_1, and 2 other mycotoxins	Maize	Accelerated solvent extraction (ASE) Elution solvent: acetonitrile:wáter (75:25 *v/v*) Clean-up with SAX Bond Elut SPE cartridges	Shiseido Capcell Pak C18 AQ column (150x4.6 mm, 5 µm) *Gradient elution:* A) Water (1% acetonitrile, 5 mM ammonium acetate) adjusted to pH 4.0 with formic acid B) Acetonitrile 1.0 mL/min	Ion trap mass analyzer (+) APCI SRM acquisition mode	9 min	50 µg/kg	[31]
Fumonisins B_1 and B_2, with 16 other mycotoxins and metabolites	Bovine milk	-Addition of sulphuric acid (pH 2.0) -Extraction and Defatting with Acetonitrile and Hexane -Clean-up with Oasis HLB SPE cartridges	Luna C18 column (150x4.6 mm, 5 µm) *Gradient elution:* A) Water:Methanol (10:90 *v/v*) with 0.02% acetic acid B) Methanol:Water (90:10 *v/v*) with 0.02% acetic acid 0.60 mL/min	Triple quadrupole mass analyzer (+) ESI SRM acquisition mode	12 min	40-50 ng/L (CCβ)	[50]
Fumonisins B_1, and B_2, with 12 other mycotoxins	Corn meal	Extraction with acetonitrile:water (75:25 *v/v*) and preconcentration with C18 SPE cartridge Clean-up with Carbograph-4 and/or Oasis HLB SPE cartridges	Alltima C18 column (250x2.1 mm, 5 µm) *Gradient elution:* A) Water (10 mM formic acid adjusted pH 3.8 with ammonia) B) Methanol (10 mM formic acid adjusted pH 3.8 with ammonia) 0.20 mL/min	Triple quadrupole mass analyzer (+) ESI SRM acquisition mode	25 min	3-5 µg/kg) (MDL)	[51]

Compounds	Samples	Sample treatment	Chromatographic conditions	Mass spectrometry conditions	Analysis time	LOQs[a]	Ref.
Fumonisins B_1, and B_2, with 9 other mycotoxins	Maize	Extraction with methanol Clean-up with AOFZDT2™ immunoaffinity column	Gemini C18 column (150x2 mm, 5 µm) *Gradient elution:* A) Water (0.5% acetic acid, 1 mM ammonium acetate) B) Methanol (0.5% acetic acid, 1 mM ammonium acetate) 0.20 mL/min	Quadrupole-ion trap (QTrap) mass analyzer (+) ESI SRM acquisition mode	50 min	0.4-1.1 µg/kg (LOD)	[52]
Fumonisins B_1, B_2 and B_3, with 21 other mycotoxins	Sweet pepper	Extraction with ethyl acetate:formica cid (99:1 *v/v*) Clean-up with SAX Bond Elut and/or C18 SPE cartridges	Gemini C18 column (150x2 mm, 5 µm) *Gradient elution:* A) Water:methanol:acetic acid (94:5:1 *v/v/v*) with 5 mM ammonium acetate B) Methanol:water:acetic acid (97:2:1 *v/v/v*) with 5 mM ammonium acetate 0.30 mL/min	Triple quadrupole mass analyzer (+) ESI SRM acquisition mode	12 min	13-27 µg/kg	[53]
Fumonisins B_1 and B_2, with 15 other mycotoxins and metabolites	Cereals	QuEChERS -10 mL water and 10 mL 0.5% acetic acid in methanol - $MgSO_4$: NaCl (4:1 *w/w*) - Defatting with hexane Accelerated solvent extraction (ASE) Elution solvent: Methanol:water:glacial acetic acid (80:19:0.5 *v/v/v*)	Zorbaz Bonus-RP column (150x2.1 mm, 3.5 µm) *Gradient elution:* A) Water (0.15% formic acid and 10 mM ammonium formate) B) Methanol (0.05% formic acid) 0.60 mL/min	Quadrupole-ion trap (QTrap) mass analyzer (+) ESI SRM acquisition mode	14 min	50 µg/kg	[38]

Table 2. (Continued)

Compounds	Samples	Sample treatment	Chromatographic conditions	Mass spectrometry conditions	Analysis time	LOQs[a]	Ref.
Fumonisin B_1 and B_2, and 2 other mycotoxins	Meat products	Extraction and defatting with water:acetonitrile:pentane -Clean-up with Oasis reversed-phase mixed anion-exchange (MAX) SPE cartridges	Gemini C_6-phenyl column (50x2 mm, 3 µm) *Gradient elution:* A) Water B) Acetonitrile 0.30 mL/min	Triple quadrupole mass analyzer (+) ESI SRM acquisition mode	8 min	150 µg/kg	[54]
Fumonisins B_1, B_2 and B_3, with 12 other mycotoxins	Beer-based drinks	QuEChERS -10 mL acetonitrile - dSPE Citrate Extraction Clean-up with InertSep C18 SPE cartridge	Acquity UPLC BEH C18 column (50x2.1 mm, 1.7 µm) *Gradient elution:* A) Water B) Methanol (2% acetic acid, 0.1 mM ammonium acetate) 0.50 mL/min	Triple quadrupole mass analyzer (+) ESI SRM acquisition mode	4 min	5 µg/L	[39]
Fumonisin B_1, with 12 other mycotoxins	Pig plasma	Deproteinization step with acetonitrile, evaporation and reconstitution with water:methanol (85:15 *v/v*)	Hypersil Gold C18 column (50x2.1 mm, 1.9 µm) *Gradient elution:* A) Water (0.1% acetic acid) B) Methanol 0.30 mL/min	Triple quadrupole mass analyzer (+) ESI SRM acquisition mode	7 min	1-2 µg/L	[55]
Fumonisin B_1 and B_2, and 8 other mycotoxins	Cereals and derived products	Extraction with acetonitrile:water (80:20 *v/v*)	Acquity UPLC BEH C18 column (100x2.1 mm, 1.7 µm) *Gradient elution:* A) Water (5 mM ammonium formate) B) Methanol 0.35 mL/min	Triple quadrupole mass analyzer (+) ESI SRM acquisition mode	4.5 min	5 µg/kg	[56]

Compounds	Samples	Sample treatment	Chromatographic conditions	Mass spectrometry conditions	Analysis time	LOQs[a]	Ref.
Fumonisins B_1, B_2 and B_3, with 11 other mycotoxins	Wines	Oasis HLB SPE extraction Multifunctional Cartridge purification (MultiSep™ #229 Ochra.)	Acquity UPLC BEH C18 column (100x2.1 mm, 1.7 μm) *Gradient elution:* A) Water B) Methanol (2% acetic acid, 0.1 mM ammonium acetate) 0.30 mL/min	Triple quadrupole mass analyzer (+) ESI SRM acquisition mode	13 min	1 μg/L	[57]
Fumonisins B_1 and B_2, with 9 other mycotoxins	Cereals	Extraction with acetonitrile:water:acetic acid (79:20:1 *v/v/v*)	C18 column (150x4.6 mm, 3 μm) *Gradient elution:* A) Water B) Methanol (0.1% acetic acid) 0.25 mL/min	Triple quadrupole mass analyzer (+) ESI SRM acquisition mode	20 min	0.1-0.5 μg/kg	[58]
Fumonisin B_1 with 8 other mycotoxins	Maize	Extraction with acetonitrile:water:acetic acid (79:20:1 *v/v/v*) Clean-up with Oasis HLB SPE cartridges	Shim-pack XR-ODS column (75x3.0 mm, 2.2 μm) *Gradient elution:* A) Water (0.1% acetic acid, 1 mM ammonium acetate) B) Methanol (0.1% acetic acid, 1 mM ammonium acetate) 0.30 mL/min	Quadrupole-ion trap (QTrap) mass analyzer (+) ESI SRM acquisition mode	16 min	2.12 μg/kg	[59]
Fumonisin B_1 with 11 other mycotoxins	Pig and human urine	Salting-out assisted LLE -Addition of 10 mL of $MgSO_4$ (2 M) -Extraction with ethyl acetate:formic acid (99:1, *v/v*) and acetonitrile:formic acid (99:1 *v/v*)	Symmetry C18 column (150x2.1 mm, 5 μm) *Gradient elution:* A) Water (0.3% formic acid, 5 mM ammonium formate) B) Methanol (0.3% formic acid, 5 mM ammonium formate) 0.25 mL/min	Triple quadrupole mass analyzer (+) ESI SRM acquisition mode	15 min	0.17 μg/L	[60]

Table 2. (Continued)

Compounds	Samples	Sample treatment	Chromatographic conditions	Mass spectrometry conditions	Analysis time	LOQs[a]	Ref.
Fumonisins B_1, B_2 and B_3, with 9 other mycotoxins	Cereals and nuts	Extraction with acetonitrile:water:acetic acid (79.5:20:0.5 *v/v/v*) Clean-up with multi-functional Inmmunoaffinity cartridges	Acquity UPLC HSS T3 RP column (100x2.1 mm, 1.7 µm) *Gradient elution:* A) Water (5 mM ammonium acetate) B) Methanol (5 mM ammonium acetate) 0.4 mL/min	Quadrupole-ion trap (QTrap) mass analyzer (+) ESI SRM acquisition mode	6 min	10 µg/kg	[61]
Fumonisins B_1, B_2 and B_3, with 14 other mycotoxins	Spices	QuEChERS -5 mL water + 5 mL methanol (1% formic acid) -2 g $MgSO_4$ + 0.5 g NaCl	Symmetry C18 column (100x2.1 mm, 3.5 µm) *Gradient elution:* A) Methanol:water (20:80 *v/v*) with 0.1% formic acid and 5 mM ammonium formate B) Methanol:water (90:10 *v/v*) with 0.1% formic acid and 5 mM ammonium formate 0.30 mL/min	Triple quadrupole mass analyzer (+) ESI SRM acquisition mode	8 min	43-88 µg/kg	[40]
Fumonisins B_1 and B_2, with 24 other mycotoxins	Finished grain and nut products	Extraction with acetonitrile:water (85:15 *v/v*)	Restek C18 column (100x2.1 mm, 3 µm) *Gradient elution:* A) Water (0.1% formic acid, 10 mM ammonium formate) B) Methanol (0.1% formic acid, 10 mM ammonium formate) 0.5 mL/min	Quadrupole-ion trap (QTrap) mass analyzer (+) ESI SRM acquisition mode	10 min	12.5 µg/kg	[62]

Compounds	Samples	Sample treatment	Chromatographic conditions	Mass spectrometry conditions	Analysis time	LOQs[a]	Ref.
Fumonisins B_1, B_2 and B_3, with 66 other mycotoxins and metabolites	Cereals, nuts and derived products	Extraction with acetonitrile:water:glacial acetic acid (20:79:1 *v/v/v*)	Gemini C18 column *Gradient elution:* A) Methanol:water:glacial acetic acid (10:89:1 *v/v/v*) with 5 mM ammonium acetate B) Methanol:water:glacial acetic acid (97:2:1 *v/v/v*) with 5 mM ammonium acetate 1.0 mL/min	Quadrupole-ion trap (QTrap) mass analyzer (+) ESI SRM acquisition mode	16 min	0.003-0.05 µg/kg (LOD)	[63]
Fumonisins B_1 and B_2, with 13 other mycotoxins	Milk thistle (*Silybum mcrianum*)	QuEChERS -8 mL of 30 mM NaH_2PO_4 (pH 7.1) -5 mL acetonitrile with 5% formic acid - 4 g $MgSO_4$, 1 g NaCl, 1 g sodium citrate, 0.5 g disodium hydrogen citrate sesquihydrate	Zorbax Eclipse C18 column (50x2.1 mm, 1.8 µm) *Gradient elution:* A) Water (0.3% formic acid, 5 mM ammonium formate) B) Methanol (0.3% formic acid, 5 mM ammonium formate) 0.40 mL/min	Triple quadrupole mass analyzer (+) ESI SRM acquisition mode	4.5 min	13.5-45.7 µg/kg	[41]
Fumonisins B_1, B_2 and B_3, with 8 other mycotoxins	Baby foods and animal feeds	Extraction with acetonitrile:water (50:50 *v/v*)	Kinetex XB-C18 column (100x2.1 mm, 2.6 µm) *Gradient elution:* A) Water (0.1% formic acid, 10 mM ammonium formate) B) Methanol (0.1% formic acid, 10 mM ammonium formate) 0.30 mL/min	Quadrupole-linear ion trap mass analyzer (+) ESI SRM acquisition mode	12 min	0.7-6.6 µg/kg	[64]

Table 2. (Continued)

Compounds	Samples	Sample treatment	Chromatographic conditions	Mass spectrometry conditions	Analysis time	LOQs[a]	Ref.
Fumonisins B_1, B_2 and B_3, with 31 other mycotoxins	Wines	Dilution with water, addition of acetonitrile (1% acetic acid), and a drying step with $MgSO_4$	Acquity UPLC BEH C18 column (100x2.1 mm, 1.7 μm) *Gradient elution:* A) Water (0.1% formic acid) B) Acetonitrile (0.1% formic acid) 0.40 mL/min	Triple quadrupole mass analyzer (+) ESI SRM acquisition mode	13 min	20 μg/L	[65]
Fumonisins B_1 and B_2, with 13 other mycotoxins	Cereals, cocoa, oil, spices, infant formula, coffee, nuts	QuEChERS -10 mL methanol:acetic acid (99.5/0.5 *v/v*) - $MgSO_4$: NaCl (4:1 *w/w*) - Defatting with hexane	Zorbax Bonus-RP C18 column (50x2.1 mm, 1.9 μm) *Gradient elution:* A) Water (0.15% formic acid, 10 mM ammonium formate) B) Methanol (0.05% formic acid) 0.50 mL/min	Quadrupole-ion trap (QTrap) and triple quadrupole mass analyzers (+) ESI SRM acquisition mode	9 min	50 μg/kg	[44]
Fumonisins B_1 and B_2, with 8 other mycotoxins	Crude extracts of nuts	Extraction with acetonitrile:water:acetic acid (79:29:1 *v/v/v*) Defatting with hexane	Hypersil GOLD C18 column (50x2.1 mm, 1.9 μm) *Gradient elution:* A) Water (1% acetic acid, 5 mM ammonium acetate) B) Methanol (1% acetic acid, 5 mM ammonium acetate) 0.50 mL/min	Triple quadrupole mass analyzer (+) ESI SRM acquisition mode	4.5 min	0.17-0.8 μg/kg	[66]
Fumonisins B_1 and B_2, with 15 other mycotoxins	Barley and malt	QuEChERS -10 mL 0.1% formic acid + 10 mL Acetonitrile -4 g $MgSO_4$+1 g NaCl	Acquity UPLC BEH C18 column (50x2.1 mm, 1.7 μm) *Gradient elution:* A) Water (0.1% formic acid) B) Methanol (0.1% formic acid, 1 mM ammonium formate) 0.40 mL/min	Triple quadrupole mass analyzer (+) ESI SRM acquisition mode	5 min	1-20 μg/kg	[42]

Compounds	Samples	Sample treatment	Chromatographic conditions	Mass spectrometry conditions	Analysis time	LOQs[a]	Ref.
Fumonisins B_1, B_2 and B_3, with 4 other mycotoxins	Dry pet food samples	Extraction with methanol:water (80:20 *v/v*) Clean-up with FumoniTest™ immunoaffinity column	Gemini C18 column (150x4.6 mm, 5 µm) *Gradient elution:* A) Water (1% acetic acid, 5 mM ammonium acetate) B) Methanol (1% acetic acid, 5 mM ammonium acetate) 0.70 mL/min	Quadrupole-ion trap (QTrap) mass analyzer (+) ESI SRM acquisition mode	-	5 µg/kg	[67]
Fumonisins B_1 and B_2, with 13 other mycotoxins	Pseudocereals, spelt and rice	QuEChERS -8 mL water + 10 mL 5% formic acid in acetonitrile -4 g $MgSO_4$, 1 g NaCl, 1 g sodium citrate, 0.5 g disodium hydrogen citrate sesquihydrate Clean-up with Immunoaffinity cartridges	Zorbax Eclipse Plus RRHD C18 column (50x2.1 mm, 1.8 µm) *Gradient elution:* A) Water (0.3% formic acid, 5 mM ammonium formate) B) Methanol (0.3% formic acid, 5 mM ammonium formate) 0.40 mL/min	Triple quadrupole mass analyzer (+) ESI SRM acquisition mode	4.5 min	0.65-1.01 µg/kg	[43]
Fumonisins B_1 and B_2, with 9 other mycotoxins	Palm kernel cake	Extraction with acetonitrile:wáter:formica cid (79:20:1 *v/v/v*)	Symmetry C18 column (150x2.0 mm, 3 µm) *Gradient elution:* A) Water (0.2% formic acid) B) Methanol 0.20 mL/min	Triple quadrupole mass analyzer (+) ESI SRM acquisition mode	16 min	18 µg/kg	[68]

[a] LOQs for Fumonisin mycotoxins.

Table 3. Selection of LC-HRMS methods for the characterization and analysis of fumonisins and/or multi-mycotoxin analysis

Compounds	Samples	Sample treatment	Chromatographic conditions	Mass spectrometry conditions	Analysis time	LOQs[a]	Ref.
28 isomers of Fumonisin B₁	Solid rice cultura	Extraction with methanol:water (75:25 v/v)	YMC-Pack J'sphere ODS H80 column (250x2.1 mm, 4 μm) *Gradient elution:* A) Water (0.1% formic acid) B) Acetonitrile (0.1% formic acid) 0.20 mL/min	TOF mass analyzer (+) ESI Full scan MS acquisition mode	35 min	-	[87]
Fumonisins B₁, B₂ and B₃, with 15 other mycotoxins	Beer	Extraction with Oasis HLB and C18 SPE cartridges	Gemini C18 column (150x2.0 mm, 5 μm) *Gradient elution:* A) Water (0.1% formic acid, 5 mM ammonium formate) B) Methanol (5 mM ammonium formate) 0.50 mL/min	Hybrid LIT-FT (LTQ Orbitrap XL) mass analyzer (+) ESI Resolving power: 100,000 FWHM Full scan MS and Full MS/dd-MS² data dependent scan acquisition modes	18 min	24-95 μg/L	[88]
Fumonisin B₁, with 8 other mycotoxins and 60 pesticides	Wines	Extraction with Oasis HLB and Bond Elut Plexa SPE cartridges	Eclipse XDB-C18 column (50x4.6 mm, 1.8 μm) *Gradient elution:* A) Water (0.1% acetic acid) B) Acetonitrile 0.50 mL/min	TOF mass analyzer (+) ESI Full-scan MS acquisition mode	17 min	2.68 μg/L	[89]
Fumonisins B₁, B₂ and hydrolyzed fumonisins	Maize-based products	Extraction with methanol:acetonitrile:citrate/phosphate buffer (25:25:50 v/v/v)	Gemini C18 column (150x2.0 mm, 5 μm) *Gradient elution:* A) Water (0.5% acetic acid) B) Methanol (0.5% acetic acid) 0.20 mL/min	Orbitrap mass analyzer (+) ESI Resolving power: 100,000 FWHM Full-scan MS and All ion fragmentation acquisition modes	30 min	10 μg/kg	[90]

Compounds	Samples	Sample treatment	Chromatographic conditions	Mass spectrometry conditions	Analysis time	LOQs[a]	Ref.
Fumonisins B_1, B_2 and B_3, with 55 other mycotoxins	Dairy products	QuEChERS -10 mL methanol:water (84:16 *v/v*) -6 g $MgSO_4$ and 1.45 g sodium acetate anhydrous -dSPE with 1.2 g $MgSO_4$, 108 mg PSA and 405 mg C18	Thermo Accucore C18 aQ column (100x2.1 mm, 2.6 µm) *Gradient elution:* A) Water (0.1% formic acid, 4 mM ammonium formate) B) Methanol (0.1% formic acid, 4 mM ammonium formate) 0.30 mL/min	Q-Orbitrap mass analyzer (+) ESI Resolving power: 17,500 FWHM Full-scan MS and full MS/dd-MS^2 data dependent scan acquisition modes	15 min	0.08-0.54 µg/kg (CCβ)	[91]

[a] LOQs for Fumonisin mycotoxins.

However, while great efforts have been made during the last years to develop more selective SPE sorbent materials, the need for robust, faster and accurate analytical methods able to determine as many mycotoxins as possible in a single analysis has also fostered the development of multiresidue strategy. In this context, QuEChERS procedure has probably become the most popular technique in many laboratories. In fact, several works can be encountered in the literature dealing with the simultaneous extraction and clean-up of a wide range of mycotoxins (including fumonisins) in cereals [38], beer-based drink [39], spices [40], milk thistle [41], barley and malt [42], among others. For instance, a QuEChERS-based sample treatment has been recently applied for the extraction and clean-up of fifteen mycotoxins in several cereal and pseudocereal samples such as buckwheat, quinoa, spelt, rice and amaranth [43]. The precision (repeatability and intermediate precision) of the developed QuEChERS-LC-MS/MS method was lower than 12% in all cases and recoveries were between 60.0 and 103.5%, fulfilling the current legislation. An alternative dispersive SPE approach have been also recently adopted by Desmarchelier et al., [44] whose optimized the QuEChERS sample preparation approach for the extraction and clean-up of 17 mycotoxins found in different matrices such as cereals, oils, infant formula, cocoa, nuts and spices. In fact, in this study the authors combined the simplicity of the QuEChERS approach with the efficiency of IAC clean-ups step, thus obtaining satisfactory recoveries, accuracy and precision. In the same way, Arroyo-Manzanares et al., [41] developed a sample treatment based on a first step using a QuEChERS-based procedure which allowed the determination of fumonisin B_1, fumonisin B_2, nivalenol, deoxynivalenol and fusarenon-X, and a subsequent clean-up based on dispersive liquid-liquid microextraction (DLLME) for the determination of several other mycotoxins (AFB$_1$, AFB$_2$, AFG$_1$, AFG$_2$, OTA, T-2, HT-2, STE, CIT and ZEN). The method which was applied to extract and seeds of milk thistle showed good recoveries between 62.3 and 98.9%, except for zearalenone in seeds samples and citrinin in extract.

LIQUID CHROMATOGRAPHY-MASS SPECTROMETRY

Nowadays, liquid chromatography-mass spectrometry (LC-MS(/MS)) is the technique of choice for the determination of fumonisins in foodstuffs. Table 1 shows a selection of published LC-MS(/MS) methods for the analysis of fumonisin mycotoxins in several kind of matrices [21-26,28-30,32-37,45-

48]. Some of these works are devoted only to the analysis of FB_1 [32, 35, 48], but most of the LC-MS(/MS) proposed methods are analyzing mainly FB_1 and FB_2 mycotoxins, and in some cases FB_3 is also targeted [21, 22, 24-26, 33, 46]. It should be mention that FB_3 standards were commercially available later than those for FB_1 and FB_2. For instance, in 2003 Vekiru et al., [45] described the determination of fumonisins and the hydrolyzed fumonisin B_1 (HFB_1) in microbial culture media by LC-ESI/MS using a single quadrupole MS analyzer, but only the quantitative determination of FB_1, FB_2 and HFB_1, with limits of detection (LOD) down to 21-53 µg/kg, was carried out. For FB_3 only a qualitative evaluation was possible because a FB_3-standard was not commercially available at the moment. The LC-MS/MS analysis of other fumonisin-derived compounds such as hydrolyzed fumonisins [28, 30, 45, 49] or N-Carboxymethyl)-fumonisin B_1 (NCM-FB_1) [28] has also been described in the literature. For instance, Seefelder et al., [28] developed an LC-ESI-MS/MS method using a triple quadrupole mass analyzer for the analysis of FB_1, HFB_1 and NCM-FB_1 in corn products. Figure 2 shows the structure of the three analyzed compounds and the chromatographic separation using a conventional C18 column.

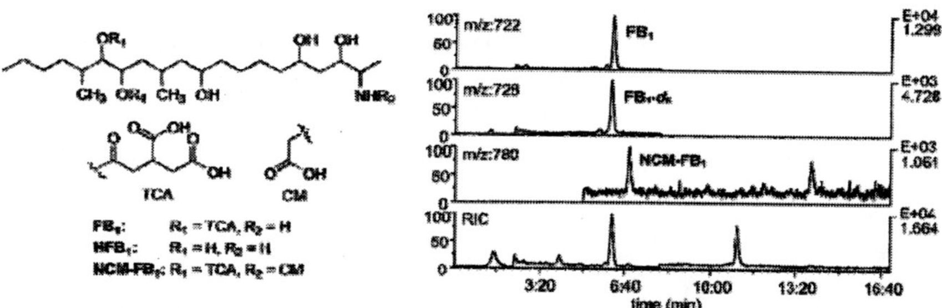

Figure 2. Structures of fumonisin B_1 (FB_1), hydrolyzed fumonisin B_1 (HB_1) and N-(carboxymethyl)fumonisin B_1 (NCM-FB_1), and LC-ESI-MS analysis of an extrusion product sample containing 45.9 ng/g of NCM-FB_1. Monitored m/z ratios were 780 (NCM-FB_1), 722 (FB_1) and 728 (FB_1-d_6). RIC: reconstructed ion chromatogram. Reprinted with permission from reference [28]. Copyright (2001) American Chemical Society.

The authors studied model experiments using corn grits spiked with FB_1 and D-glucose or sucrose to determine the significance of NCM-FB_1 formation relating to the fate of FB_1 during the extrusion process. The results clearly showed the low significance of NCM-FB_1 as a reaction product of FB_1 and reducing sugars in heat-treated food, so the fate of FB_1 in heat-treated corn products cannot be explained with the formation of NCM-FB$_1$. However, this was the first report of a screening for NCM-FB$_1$ in corn-containing products from the German market.

Gazzotti et al., [30] analyzed fumonisins B$_1$ and B$_2$, together with their hydrolyzed metabolits HFB$_1$ and HFB$_2$, in pig liver by reversed-phase LC-MS/MS using a triple quadruple mass analyzer (Figure 3). The method was based on an easy extraction procedure followed by SPE purification, showing good performance characteristics and being sensitive enough for this kind of analysis (LOQ of 10 µg/kg). The authors applied the method to the analysis of seven pig liver samples to test its applicability to evaluate exposure of food animals to fumonisins. The results showed an evident contamination with FB$_1$ in almost all samples analyzed (values from traces to 42.5 µg/kg), and only HFB$_1$ was detected in one sample (17.3 µg/kg). Only traces of FB$_2$ were detected in 5 samples while no presence of HFB$_2$ was observed in any sample.

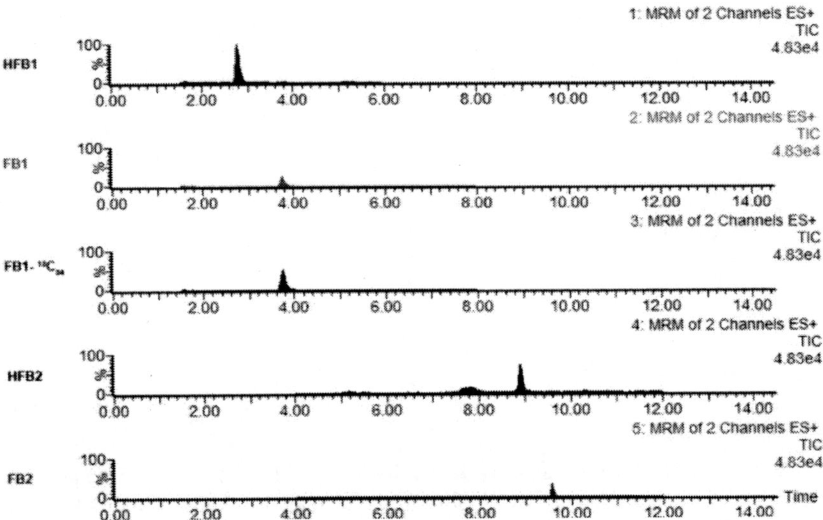

Figure 3. Total ion current (TIC) chromatograms of blank sample of a wild bred pig liver fortified at 10 ng/g of each analyte and 30 ng/g of internal standard. Reprinted with permission from reference [30]. Copyright (2011) Elsevier.

But in general, most of the authors are proposing multi-mycotoxin LC-MS(/MS) analytical methods. Table 2 shows a selection of published LC-MS(/MS) methods for the multi-residue analysis of mycotoxins, including fumonisis, in several kind of matrices [31, 38-44, 50-68]. Again, most of the multi-mycotoxin LC-MS(/MS) methods are including the analysis of the three most important fumonisin mycotoxins (FB_1, FB_2 and FB_3) or at least two of them (FB_1 and FB_2), and only few works are considering only FB_1 [31, 55, 59, 60]. For instance, Song et al., [60] developed and applied a salting-out assisted liquid-liquid extraction (LLE) method for the multi-mycotoxin biomarkers analysis in pig urine with liquid chromatography-tandem mass spectrometry using a triple quadrupole mass analyzer. The direct determination of urinary mycotoxins could be a better approach to assess individual's exposure than the indirect estimation from average dietary intakes. By using a conventional reversed-phase C18 column the chromatographic separation of 12 mycotoxins, including FB_1, was achieved in less than 15 minutes (Figure 4). The method was also applied to the analysis of these mycotoxins in human urine.

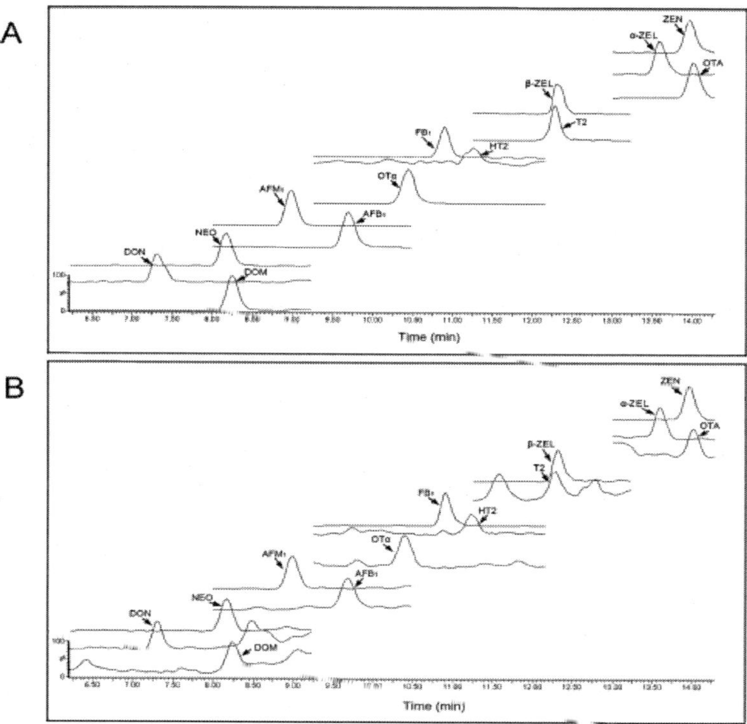

Figure 4. Chromatograms of 12 mycotoxins in a standard solution (A) and in pig urine after salting-out assisted liquid-liquid extraction (B). The concentration of each compound is as follows: deoxynivalenol (DON), 5 ng/mL; neosolaniol (NEO), 2 ng/mL; aflatoxin M_1 (AFM_1), 1 mg/mL; aflatoxin B_1 (AFB_1), 0.2 ng/mL; Ochratoxin α (OTα), 2.5 ng/mL; fumonisin B_1 (FB_1), 0.5 ng/mL; HT2 toxin (HT2), 0.5 ng/mL; T2 toxin (T2), 0.2 ng/mL; β-zearalenol (β-ZEL), 1 ng/mL; α-zearalenol (α-ZEL), 1 ng/mL; zearalenone (ZEN), 1 ng/mL; and ochratoxin A (OTA), 0.1 ng/mL. Reprinted with permission from reference [60]. Copyright (2013) Elsevier.

The authors reported extraction recoveries in the range of 70-108%, with intra-day relative standard deviations (RSD) and inter-day RSD values lower than 25% for most of the compounds at three different concentration levels. Quantification was carried out by matrix-matched calibration to correct matrix effects. Regarding FB_1 a limit of quantification (LOQ) value of 0.17 µg/L was achieved.

Chromatographic Separation

As can be seen in Tables 1 and 2, chromatographic trends for the LC-MS(/MS) analysis of fumonisins, and mycotoxins in general, are based on reversed-phase separations with C18 columns and using acidified water and methanol or acetonitrile solvents as mobile phases. In general, formic acid and acetic acid, as well as their ammonium salts, are used as mobile phase modifiers to control separation pH. However, some of the first publications dealing with the analysis of fumonisins also described the use of trifluoroacetic acid (TFA) for that purpose [28, 32], although it was soon removed because of the well known chromatographic contamination problems associated with TFA. For instance, in 2002 Seefelder et al., [32] described the use of water and acetonitrile, both with 0.05% TFA, as mobile phases for the LC-MS analysis (using a single quadrupole mass instrument) of FB_1 in asparagus spears and garlic bulbs from Germany, achieving limits of detection (LODs) in the range of 26-36 µg/kg. FB_1 was found in the four garlic samples analyzed at concentrations ranging from 26 to 94 µg/kg. All samples contained also lower concentrations of FB_2 and FB_3 although accurate quantification was not possible because of the lack of standards at the moment.

In general, gradient elution is being used to achieve the chromatographic separation of mycotoxins, and only some works dealing with the specific analysis of fumonisins (Table 1) are proposing the use of isocratic elution modes [21, 22, 33, 35, 47]. For instance, Gazzotti et al., [35] determined fumonisin B_1 in bovine milk by LC-MS/MS with a conventional C18 chromatographic column under isocratic elution using a mixture of 75% of water:acetonitrile (90:10 *v/v*) with 0.3% formic acid and 25% of acetonitrile (0.3% formic acid) as mobile phase. With a mobile phase flow-rate of 0.3 mL/min, FB_1 was determined in less than 4 minutes (Figure 5), with a LOQ value of 0.1 µg/kg. The authors analyzed 10 commercial milk samples and the presence of FB_1 was detected and quantified in 8 of them (at 0.26-0.43 µg/kg levels).

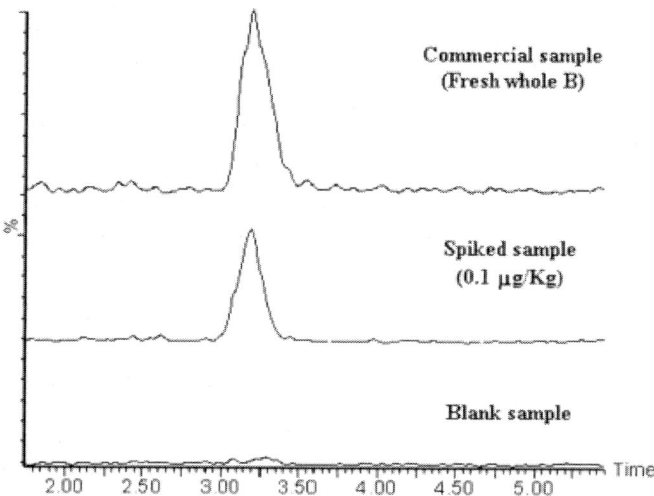

Figure 5. LC-ESI-MS/MS chromatograms of blank, spiked at 0.1 µg/kg and naturally contaminated milk obtained in positive ESI mode (722.15>334.10 and 722.15>352.10 m/z). Reprinted with permission from reference [35]. Copyright (2009) Elsevier.

In another example, Paepens et al., [21] developed a reversed-phase C18 LC-MS/MS method for the analysis of FB_1, FB_2 and FB_3 in cornflakes also under isocratic conditions by using acetonitrile:water (60:40 v/v) with 0.3% formic acid solution as mobile phase. By using a mobile phase flow-rate of 0.3 mL/min, chromatographic separation of all three fumonisins was achieved in less than 8 minutes, with LOQ values in the range 15-40 µg/kg and coefficients of variation (CVs) under repeatability conditions lower than 14%.

Almost all published LC-MS(/MS) methods dealing with the specific analysis of fumonisin and fumonisin-related compounds are using conventional 5 µm particle size columns (Table 1), and other works conventional 3-3.5 µm particle size columns [29, 37, 47, 48]. This trend is also observed for several multi-mycotoxin LC-MS(/MS) methods (Table 2) [31, 38, 40, 50-54, 58, 60, 62, 67, 68]. In general, relatively long analysis times are usually reported with the use of these columns. For instance, the LC-MS analysis of FB_1, FB_2 and HFB_1 carried out by Vekiru et al., [45] required 18 minutes using a 5 µm particle size C8 chromatographic column. Zitomer et al., [24] reported an LC-MS/MS method for the analysis of FB_1, FB_2 and FB_3 with a chromatographic run of 20 minutes using a conventional 5 µm particle size C18 column. However, in the commented work the analysis of fumonosins was performed together with biomarkers of disrupted sphingolipid metabolism such as free sphingoid bases and sphingoid base 1-phosphates in tissues of maize seedlings, which obviously required a longer chromatographic separation, while with the proposed method the three fumonisins eluted within the first 7 minutes. The analysis of these compounds is interesting because fumonisins of the B-series are known to disrupt sphingolipid metabolism in both plants and animals via inhibition of ceramide synthase [69, 70], resulting in the accumulation of free sphingoid bases and their 1-phosphates [70]. These compounds could be signaling molecules in plants, as well as inducers of both increased proliferation and apoptosis. Recently, it was described that fumonisins are involved in the development of leaf lesions in the sweet maize line "Silver Queen" [71]. So, the ability to quickly analyze both fumonisin and fumonisin-induced disruption of sphingolipid metabolism is then useful for determining not only the presence of fumonisins in tissues but also their biological effects.

As previously commented, conventional 3-5 µm particle size columns have also been employed in LC-MS/MS multi-mycotoxin analysis, reporting in some cases very long analysis times. For instance, Lattanzio et al., [52] developed a LC-MS/MS method for the simultaneous determination of aflatoxins (B_1, B_2, G_1, G_2), ochratoxin A, fumonisins (B_1, B_2), deoxynivalenol, zearalenone, and T-2 and HT-2 toxins in maize by using a conventional 5 µm particle size C18 reversed phase column. In this work, a quadrupole-ion trap (QTrap) mass analyzer with positive/negative switching ESI mode was employed. Baseline separation of the 12 analyzed mycotoxins was achieved within a 50 minutes chromatographic run. Obviously, a very good chromatographic separation is required if both positive and negative ESI modes are intended to be used in the same run, which is probably the reason of this long reported analysis time (together with a very low mobile phase flow-rate, 0.2 mL/min). In contrast, Desmarchelier et al., [38] described a multi-residue method for the analysis of 17 mycotoxins (including FB_1 and FB_2) in cereals by LC-ESI-MS/MS using also a QTrap mass analyzer and positive/negative switching ESI mode. But in this case chromatographic separation (Figure 6) of 17 mycotoxins (including FB_1 and FB_2) was achieved in less than 14 minutes, which can be probably related to fact that a 3.5 µm particle size C18 column was used, being able to achieve better chromatographic efficiency, together with a higher mobile phase flow-rate (0.6 mL/min). LOQs in the range 0.5-100 µg/kg (for fumonisins 50 µg/kg) were achieved. The authors evaluated the trueness of the proposed method through the participation in four proficiency tests and by the analysis of two certified reference materials and one quality control material. Satisfactory trueness values (73-130%) were obtained.

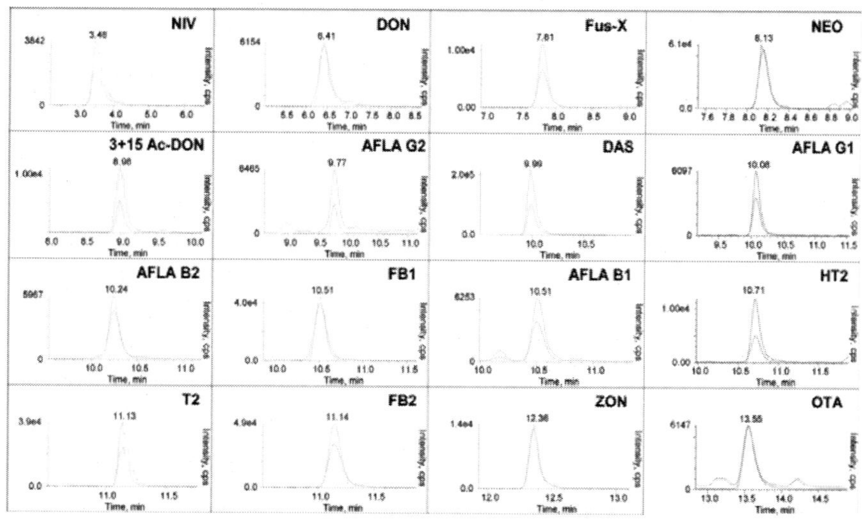

Figure 6. LC-ESI-MS/MS chromatograms of mycotoxins from an extract of oat flour. Reprinted with permission from reference [38]. Copyright (2010) American Chemical Society.

But today there is an increase need for applications in food safety analysis able to cope with a large number of analytes in very complex matrices [72, 73]. Laboratories, especially regarding food safety analysis, are demanding cost-effective methodologies with reduced analysis time. One modern approach in LC methods which enable the reduction of the analysis time without compromising resolution and separation efficiency is the use of ultrahigh pressure liquid chromatography (UHPLC) methods either using sub-2 μm [74, 75] or porous shell columns (with sub-3 μm superficially porous particles) [76-78]. Thus, the use UHPLC methods employing sub-2 μm columns has been highly reported for the analysis of mycotoxins [39, 41-44, 55-57, 59, 61, 65, 66], and also for the specific analysis of fumonisins [46]. These methods allowed a considerable reduction in chromatographic analysis times. For instance, Ren et al., [46] achieved the UHPLC-MS/MS analysis of fumonisins FB_1, FB_2 and FB_3 in less than 4 minutes in maize samples, with LOQ values between 0.4 and 1.65 μg/kg with good recoveries (81-97%) and good precision (RDS values lower than 11%).

But nowadays, the use of sub-2μm particle size columns is becoming very popular for the multimycotoxin analysis. For instance, Tamura et al., [39] described the analysis of 15 mycotoxins in less than 4 minutes by UHPLC-ESI-MS/MS using an Acquity UPLC BEH C18 column (50x2.1 mm, 1.7 μm

 Paolo Lucci, Rocío Castro-Ríos and Oscar Núñez

particle size). In fact, although the analysis of almost all mycotoxins was carried out in negative ESI mode, the positive ESI analysis of fumonisins B_1, B_2 and B_3 was achieved in less than 1.5 minutes, obtaining LOQ values down to 5 µg/kg with good repeatability (RSD values lower than 14.6%, in general). Arroyo-Manzanares et al., [41] also developed an UHPLC-ESI-MS/MS method able to analyze 15 mycotoxins (including FB_1 and FB_2) in Milk thistle (*Silybum marianum*) in less than 4.5 minutes by employing a similar C18 column (Zorbax Eclipse C18 column, 50x2.1 mm, 1.8 µm particle size). Recently, the same group applied this method to the analysis of these mycotoxins in pseudocereals, spelt and rice (see an example of the chromatographic separation in Figure 7) [43].

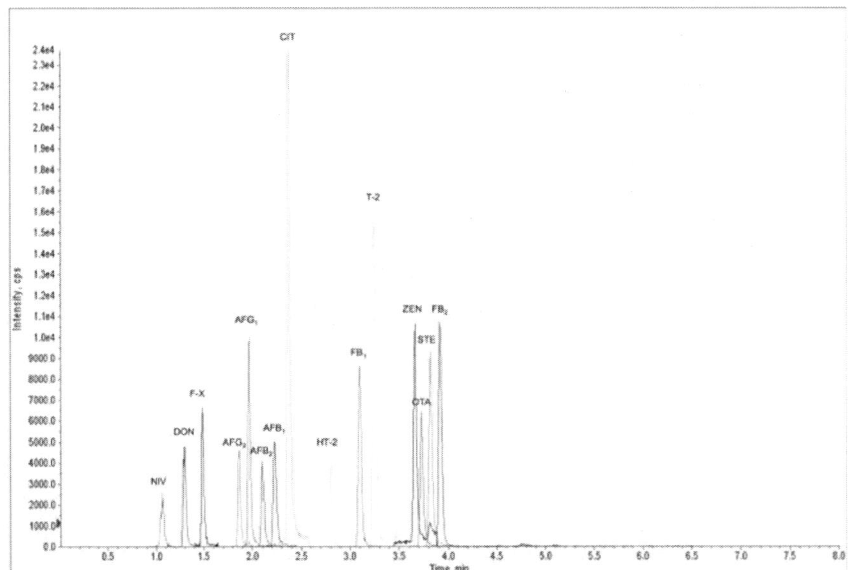

Figure 7. UHPLC-ESI-MS/MS chromatogram of a spiked white rice sample. Reprinted with permission from reference [43]. Copyright (2014) Elsevier.

In both works the authors employed a QuEChERS method for the extraction of the analyzed mycotoxins. By using a triple quadrupole mass analyzer and selected reaction monitoring (SRM) acquisition mode, LOQ values for fumonisins in the 13.5-45.7 µg/kg range for milk thistle analysis [41] and in the 0.65-1.01 µg/kg range in the pseudocereals, spelt and rice analysis [43] were obtained.

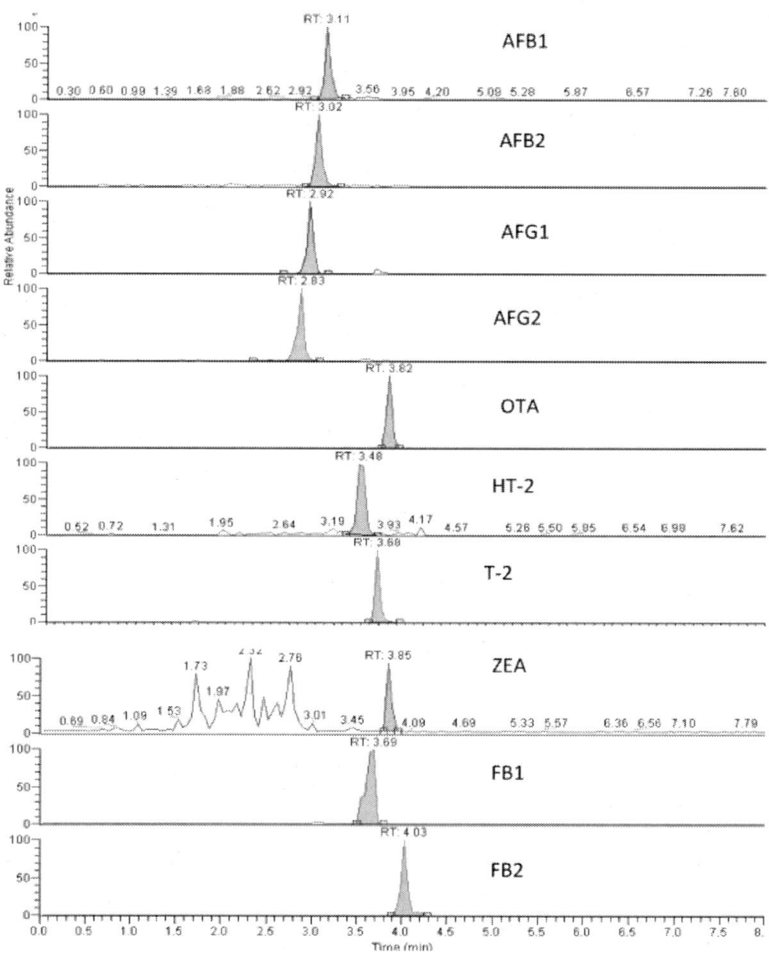

Figure 8. UHPLC-ESI-MS/MS chromatogram of a spiked crude extract of walnut with 10 mycotoxins. Reprinted with permission from reference [66]. Copyright (2014) Elsevier.

Recently, Skrbic et al., [66] also reported the determination of 10 mycotoxins (including FB_1 and FB_2) in crude extracts of nuts with a very short analysis time (4.5 min) by employing UHPLC-ESI-MS/MS and a Hypersil Gold C18 column (50x2.1 mm, 1.9 μm). This was a very simple method that only employed a simple extraction with an acetonitrile:water:acetic acid solution and defatting with hexane. As an example, Figure 8 shows the chromatogram of a mixture of all analyzed mycotoxins spiked in a crude extract of walnut at the lowest calibration level employed for quantification.

With this method, LOQ values for fumonisin analysis down to 0.17-0.8 μg/kg were obtained, with very good recoveries in a variety of nuts matrices (71-140%) and good precision (RSD values lower than 12%). In another recent example, Pizzutti et al., [65] reported the determination of 34 mycotoxins (including FB_1, FB_2 and FB_3) in wine samples by UHPLC-ESI-MS/MS using a longer Acquity UPLC BEH C18 column (100x2.1 mm, 1.7 μm particle size), and achieving a total analysis time of 13 minutes.

Fast chromatographic and high efficiency separations can also be achieved using columns packed with superficially porous particles, as previously commented, also known as porous shell or fused-core columns [72, 78]. The use of this kind of particles was first reported in 1960s with the objective of reducing analyte diffusion distance to minimize mass transfer in chromatographic columns. Today, these columns are commercially available under several brand names, and usually consist of a silica particle 1.7-1.9 μm fused-core and 0.5-0.35 μm (respectively) layer of porous silica coating, creating a total particle diameter of 2.7-2.6 μm (respectively) depending on the brand. These particles exhibit efficiencies that are comparable to sub-2μm porous particles, but with modest backpressures, that may be due to the narrow particle size distribution and higher density of fused-core particles. Further, the small diffusion path for the analyte may reduce the resistance to mass transfer in the chromatographic column, thus allowing the operation at higher flow-rates with minimal losses in efficiency [78]. These porous shell columns have also been employed for the analysis of fumonisins [25, 64]. For instance, Zhang et al., [64] developed an LC-ESI-MS/MS method using a fused-core kinetex XB-C18 column (100x2.1 mm, 2.6 μm) for the analysis of mycotoxins (including FB_1, FB_2 and FB_3) in baby foods and animal feeds. The proposed method provided sufficient selectivity, sensitivity, accuracy, and reproducibility to screen for aflatoxins at ng/g concentrations and deoxynivalenol and fumonisins and low μg/g concentrations levels.

Although most of the developed LC-MS/MS methods for the analysis of fumonisins and mycotoxins in general are based on C18 chromatographic column (Tables 1 and 2), other stationary phases such as C8 [45], HSS T3 silica-based bonded phase [61], C_6-phenyl phases [54], as well as perfluorinated-based phases [25] have also been employed. For instance, Sorensen et al., [54] proposed the use of a Gemini C_6-phenyl column (50x2 mm, 3 μm particle size) for the LC-MS/MS analysis of ochratoxin A, mycophenolic acid, and fumonisins B_1 and B_2 in meat products, which might be one of the most challenging matrices. With a simple extraction and defatting using a water:acetonitrile:pentane solution, and a clean-up step with

Oasis reversed-phase mixed anion-exchange SPE cartridges, the proposed LC-MS/MS method was capable of detecting as low as 6 µg/kg of FB_2 in this challenging matrix, with a very good precision and accuracy.

Vaclavikova et al., [61] proposed the use of an HSS T3 silica-based bonded phase, that are specially designed for the analysis of polar and non-polar compounds and 100% compatible with aqueous mobile phases, for the UHPLC-ESI-MS/MS analysis of 12 mycotoxins (including FB_1, FB_2 and FB_3) in cereals and nuts (see, as an example, the chromatographic separation in Figure 9). The authors achieved the separation of these mycotosins within a chromatographic run of 6 minutes. LOQ values for fumonisins of 10 µg/kg were obtained. The method trueness was tested with analytical spikes and certified reference materials, with recoveries ranging from 71% to 112% for all of the targeted mycotoxins.

Other stationary phases complementary to the alkyl-type (C8 and C18 phases) are the fluorinated reversed ones [72, 79]. Two types of highly fluorinated siloxane-bonded stationary phases can be distinguished, perfluoroalkyl and pentafluorophenyl, showing different separation characteristics. Perfluoroalkyl ones exhibit enhanced retention and selectivity for the separation of halogenated compounds and shape selectivity for the separation of positional isomers and non-planar molecules but this type of stationary phases are rarely used in food analysis. However, the pentafluorophenyl (PFP) stationary phases are more hydrophobic and display higher phase selectivity.

Compared to traditional alkyl-type stationary phases which achieved selectivity based on hydrophobic interactions, the PFP stationary phases use multiple retention mechanisms such as dipole-dipole, π-π, and dispersion interactions in addition to hydrophobic interactions. Due to its unique selectivity and the higher retention observed for polar compounds the use of these columns is becoming popular in food analysis, and it has also been employed in the analysis of mycotoxins [25]. For instance, Bryla et al., [25] proposed the use of a porous shell Kinetex PFP column (100x2.1 mm, 2.6 µm particle size) for the LC-MS/MS determination of fumonisins B_1, B_2 and B_3 in 49 cereals (42 maize-based and 7 wheat-bsed products). The authors were able to analyze these compounds with a LOQ value of 25 µg/kg. Unfortunately, the authors did not take advantage of the porous shell column properties to reduce the achieved chromatographic run (26 min) by working under UHPLC conditions.

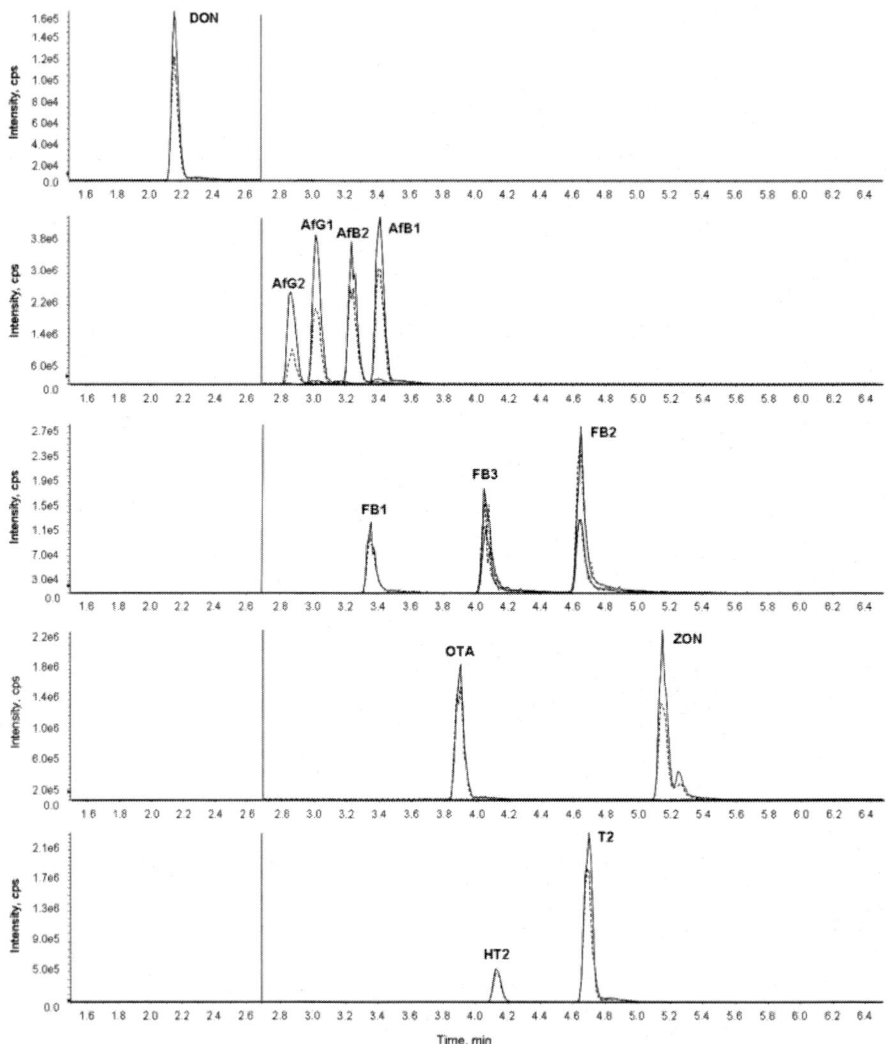

Figure 9. UHPLC-ESI-MS/MS chromatogram of mycotoxins using a HSS T3 silica-bonded phase chromatographic column. Reprinted with permission from reference [61]. Copyright (2013) Elsevier.

Ionization Modes and Mass Analyzer Instruments

Among the atmospheric pressure ionization (API) techniques frequently used with LC-MS(/MS) methods, electrospray ionization (ESI) mode is the preferred one when dealing with the analysis of fumonisins, as can be seen in

Table 1, and always in positive ionization mode. As an example, Ren et al., [46] studied both positive and negative electrospray ionization of fumonisins B_1, B_2 and B_3 in acetonitrile:water (1:1 v/v) solutions with a triple quadrupole mass spectrometer, and Figure 10 shows the mass spectra obtained. As can be seen, the results indicated that all three fumonisins dissolved in acetonitrile-water (1:1 v/v) solution could generate the $[M+H]^+$ precursor ions with higher abundances under the (+)ESI mode than with (-)ESI.

Electrospray is also frequently selected when dealing with multimycotoxin LC-MS(/MS) analysis (Table 2), although other API techniques have also been selected. For instance, Royer et al., [31] described the quantitative analysis of some Fusarium mycotoxins (deoxynivalenol, zearalenone and fumonisin B_1) in maize samples using accelerated solvent extraction before LC-APCI-MS/MS. The results obtained by the authors showed that enhanced MS response was observed in both positive and negative APCI for deoxynivalenol and zearalenone, while better response was achieved with positive ESI mode for FB_1. However, the authors selected APCI for the quantitative analysis of all three mycotoxins because (i) the sensitivity achieved for FB_1 was within the detection limit fixed by several authorities and (ii) because it was known that APCI minimized matrix effects.

Several mass spectrometers have been described for the LC-MS(/MS) analysis of fumonisins (Table 1) or the multi-residue analysis of mycotoxins (Table 2). As can be seen, very few works are employing single quadrupole mass analyzers to propose LC-MS methods [32, 45]. The typical trend when dealing with mycotoxins analysis (including fumonisins) is to use triple quadrupole mass spectrometers because of their enhanced sensitivity compared to single quadrupole or ion trap mass analyzers. For example, limits of detection down to 0.01-0.01 µg/kg for FB_1 and FB_2 in food and feed samples were achieved with a triple quadrupole mass analyzer by Khayoon et al., [29], or in the range 0.17-0.33 µg/L for fumonisin B_1 in pig and human urine by Song et al., [60] when dealing with the multimycotoxin analysis with other 11 mycotoxins by LC-MS/MS also with a triple quadrupole mass spectrometer.

Mass spectrometers based on ion trap (IT) technology such as conventional ion trap mass analyzers [25], linear ion trap mass analyzers [24] or hybrid quadrupole-linear ion trap (QTrap) mass analyzers [23, 37, 38, 44, 48, 59, 62, 64] have also been described for the analysis of fumonisins and mycotoxins in general. For instance, Zitomer et al., [24] described the analysis of fumonisins and biomarkers of disrupted sphingolipid metabolism in tissues of maize seedlings by LC-MS/MS using a linear ion trap and achieving LODs of 10 µg/kg for fumonisins.

A
FB1-N 57 (0.792) Cm (65:71) 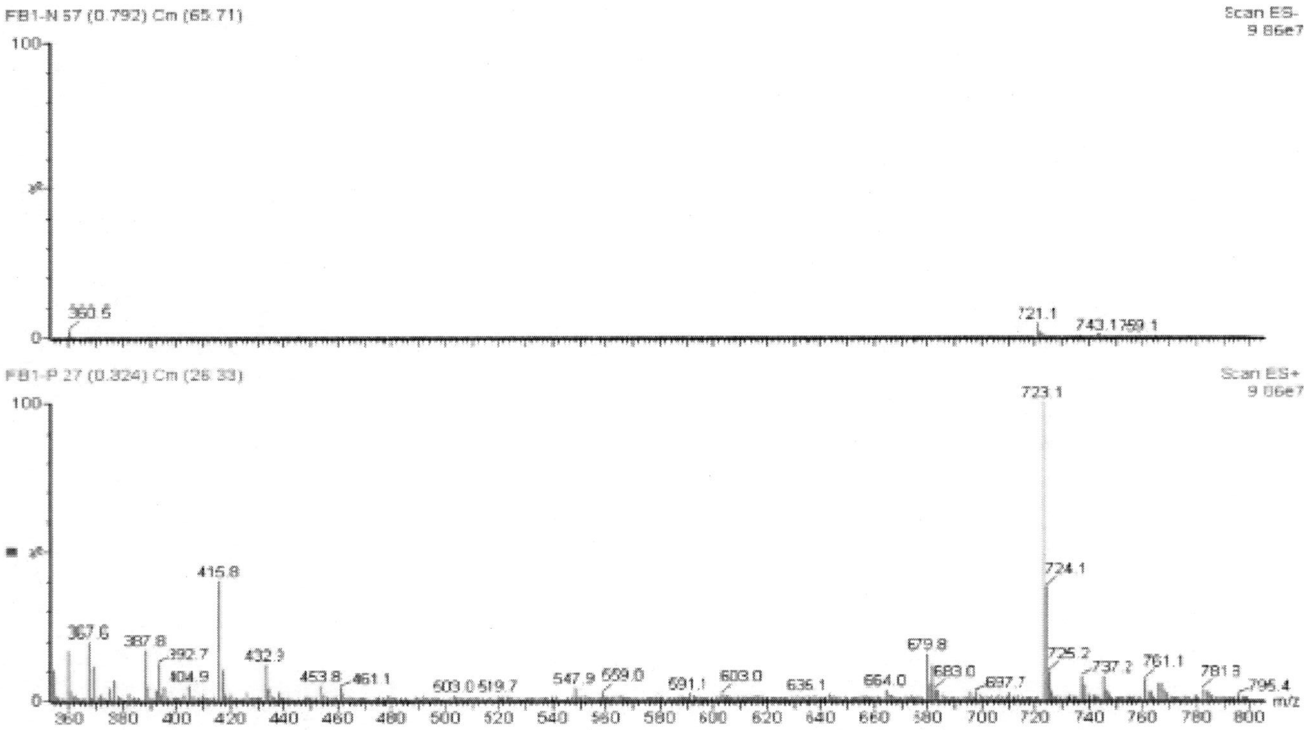
Scan ES-
9.86e7
100
%
0
360.5
721.1
743.1 769.1
FB1-P 27 (0.324) Cm (26:33)
Scan ES+
9.06e7
100
%
0
367.6
387.8
392.7
404.9
415.8
432.3
453.8
461.1
603.0 519.7
547.9 559.0
591.1
603.0
636.1
664.0
679.8
683.0
697.7
723.1
724.1
725.2
737.2
761.1
781.3
795.4
m/z
360 380 400 420 440 460 480 500 520 540 560 580 600 620 640 660 680 700 720 740 760 780 800

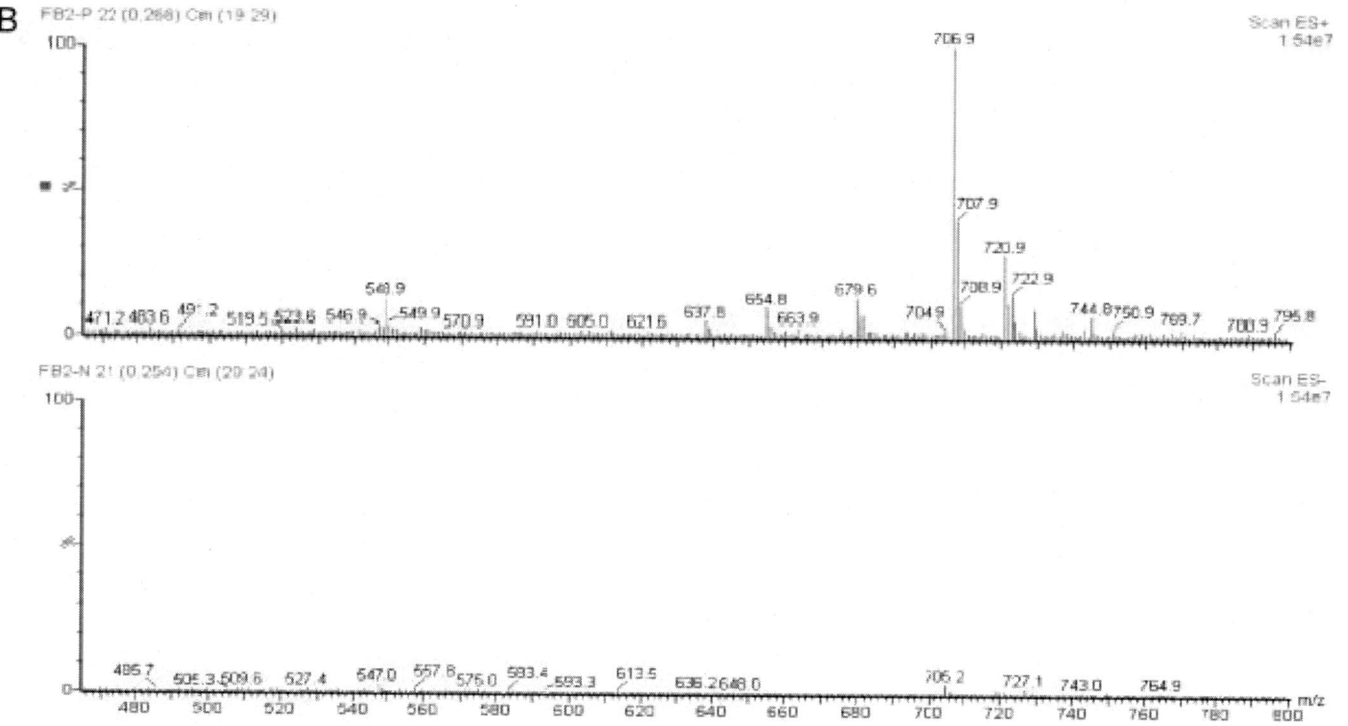

Figure 10. (Continued)

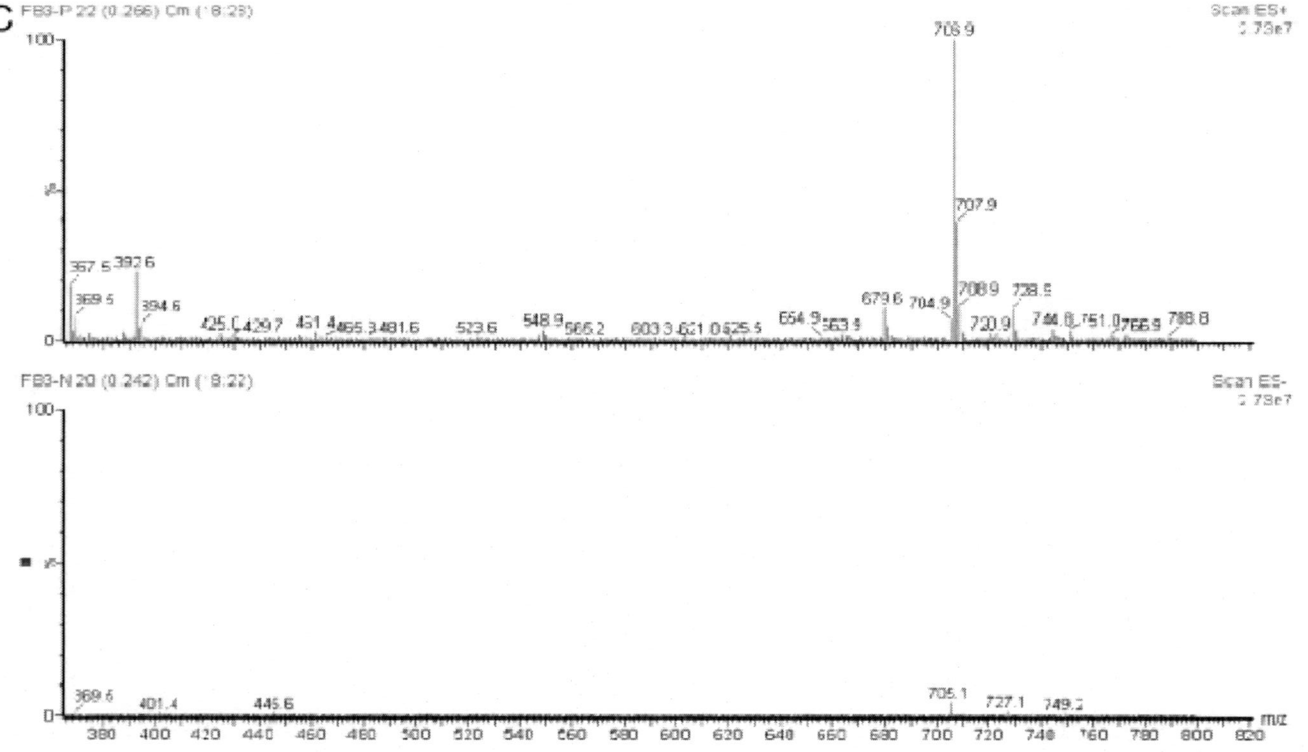

Figure 10. Mass spectra of (A) FB$_1$, (B) FB$_2$ and (C) FB$_3$ under (+)ESI (top) and (-)ESI (bottom) ionization modes. Reprinted with permission from reference [46]. Copyright (2011) Elsevier.

However, because of the high sensitivity also achieved with the new hybrid quadrupole-linear ion trap (QTrap) mass instruments available today, in some cases even better than the one achieved by some triple quadrupole instruments, these QTrap mass analyzers are becoming also very popular for the LC-MS/MS analysis of mycotoxins.

For instance, LODs as low as 0.01-0.04 µg/kg for fumonisins B_1 and B_2 were achieved by Khayoon et al., [29] by LC-ESI-MS/MS using a QTrap mass analyzer. Nevertheless, it should be mention that in spite of the different acquisition modes available when using this kind of instruments, such as the monitoring of product ion scan spectra on MS^2 experiments, all the described works in Tables 1 and 2 dealing with the LC-MS/MS analysis of fumonisins, or mycotoxins, are acquiring MS data in selected reaction monitoring (SRM) mode, implying a loss on structural information for obtaining higher sensitivity.

An interesting work dealing with the use ion trap mass analyzers in this field is the one described by Bartók et al., [49] in 2006. In this work, the authors detected new fumonisin mycotoxins and fumonisin-like compounds by reversed-phase LC-ESI-MS^2 using an ion trap mass analyzer working in product ion scan acquisition mode. For that purpose, the authors produced fumonisins in a rice culture infected with *Fusarium verticillioides* and, to decrease the possibility of the formation of artifacts, fumonisins were analyzed by the proposed LC-ESI- MS^2 method immediately after the extraction of the culture material without any sample clean-up. On the basis of the IT-MS^2 data, detailed fragmentation pathways including new mechanisms were proposed for the different series of fumonisins. In addition to already known fumonisins, the authors detected numerous new fumonisin mycotoxins and fumonisin-like compounds.

Confirmation and Quantification Aspects

As previously commented, triple quadrupole mass analyzers, and also hybrid quadrupole-linear ion trap mass analyzers, are the most popular instruments for the LC-MS/MS analysis of fumonisins, and other mycotoxins, due to their higher sensitivity and selectivity, especially when operated in SRM acquisition mode. This is the preferred acquisition mode in order to achieve the confirmation criteria established by the European Union. For confirmation of the identity of the analytes, the EU directive 2002/657/EC [80] established that two SRM transitions must be monitored to comply with a system of required identification points when low resolution LC-MS/MS is used. In addition, the deviation of the relative intensity of the recorded transitions must not exceed certain percentage of that observed with reference standards, and the retention time must not deviate more than 2.5%. To achieve this, and when working with (+)ESI mode, the $[M+H]^+$ ion is selected as precursor for each fumonisin (m/z values of 723, 707 and 707 for FB_1, FB_2 and FB_3, respectively), and two fragment ions are monitored, the most abundant one as the quantitative fragment ion, and the second most abundant one as the qualitative fragment ion (for confirmation purposes). The fragmentation pathway of fumonisins consisted of losses of water and the tricarboxylic acid (TCA) side chains from the alkyl backbone (see Figure 1) resulting in the fragments $[M+H-H_2O]^+$, $[M+H-2H_2O]^+$, $[M+H-TCA]^+$, $[M+H-TCA-H_2O]^+$, $[M+H-2TCA]^+$, $[M+H-2TCA-H_2O]^+$ and $[M+H-2TCA-2H_2O]^+$, that for fumonisin B_1 will produce m/z values at 705, 687, 546, 528, 370, 352 and 334, respectively [81]. The two selected product ions vary from work to work according to the fragmentation conditions used.

The determination of fumonisins is quite difficult regarding quantification aspects. An interesting work reported by Dall'Asta et al., [82] showed the difficulties in fumonisin determination due to the presence of hidden fumonisins. The authors compared the results obtained by five independent methods for the quantification of fumonisins B_1, B_2 and B_3 in raw maize. Five naturally contaminated maize samples and a reference material were analyzed in three different laboratories. Although each method was validated and common calibrants were used, a poor agreement about fumonisin contamination levels was obtained. In order to investigate the interactions among analytes and matrix leading to this lack of consistency, the occurrence of fumonisin derivatives was checked by the authors. Significant amounts of hidden fumonisins were detected for all the considered samples. Since different methods were used in different laboratories, the authors cannot differentiate between method and laboratories bias. However, each method was thoroughly in-house validated, and the same calibrants were used in all laboratories. Sample stability and homogeneity was also checked by including a reference material in the considered sample set. Thus, we can assume that laboratory bias was not the reason for the different measured fumonisin levels. These unexpected results pointed out a serious analytical problem in fumonisin determination.

One of the biggest problems when dealing with the quantification of fumonisins, and other mycotoxins, is matrix effect. For this reason, and because quantification by standard addition would be very time-consuming and expensive, many authors propose the use of matrix-matched calibration for quantification purposes. However, the use of isotopic dilution by employing labeled internal standards has also shown to be a good choice for the quantification of mycotoxins [48, 64, 81]. For instance, Zhang et al., [64] studied the slope ratios between matrix-matched calibration curves and solvent calibration curves to graphically suggest matrix effects in the test samples (grain, oat meal, rice, dog food, animal feed and cat food) used, and the results are shown in Figure 11. If the suppression or enhancement produced by the matrix is marginal, the ratio (matrix-matched/solvent-only standard) would equal or be very close to 1.00; if there is severe suppression/enhancement, the ratio would deviate from one.

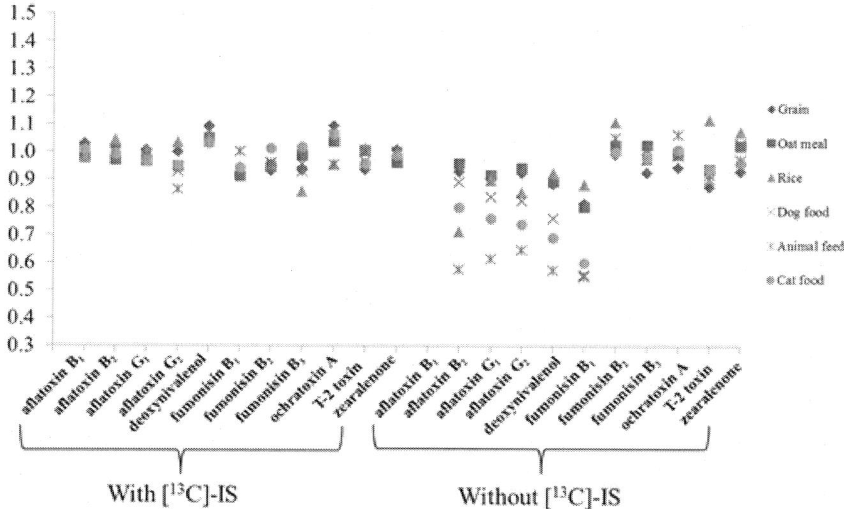

Figure 11. Slope ratios of matrix-matched and solvent calibration curves used to suggest matrix effects in baby food and animal feed matrices. Reprinted with permission from reference [64]. Copyright (2014) American Chemical Society.

By using isotopic dilution employing ^{13}C-uniformly labeled mycotoxins as internal standards, the ratios of the 11 mycotoxins in the six tested matrices by the authors center toward 1.00, ranging from 0.86 (aflatoxin G_2 in an animal feed extract) to 1.09 (ochratoxin A in a grain-based baby food extract) as can be seen in Figure 11. This is because target mycotoxins and the ^{13}C-labeled ones undergo the same ionization conditions, and therefore the co-eluted matrix components causing suppression or enhancement have the same impact on target mycotoxins and its ^{13}C-labeled internal standards during ionization.

Therefore, if a calibration curve is established using relative responses of the mycotoxins and ^{13}C-labeled internal standard, the matrix effect would be corrected.

LIQUID CHROMATOGRAPHY-HIGH RESOLUTION MASS SPECTROMETRY

In the last years, the use of liquid chromatography-high resolution mass spectrometry-based methods (LC-HRMS) for the analysis of mycotoxins has increased [83]. HRMS enables accurate mass measurements at high resolving power and has recently evolved to the stage that is rapidly causing a shift from unit-resolution quadrupole-dominated instrumentation. Different HRMS configurations are currently available, using single analyzers (Orbitrap or time-of-flight, TOF) and hybrid configurations (quadrupole coupled to Orbitrap, ion trap coupled to Orbitrap, quadrupole coupled to TOF, etc.). With these devices compound's or fragment's mass can be measured with sufficiently high accuracy that its elemental composition can be determined directly. Accurate mass and isotopic pattern are used for confirming the identity of a compound or even identification of an unknown [84].

In contrast with low resolution techniques, LC-HRMS can differentiate between two compounds with the same nominal mass-to-charge ratio (m/z). Therefore, chromatographic separation of such compounds is not required, unless matrix effects are observed. In LC-MS/MS mainly target compounds are analyzed, although it can be used for untargeted compound analysis if working in the full-scan MS mode. Nevertheless, full-scan analysis reduces analytical sensitivity compared with single ion monitoring (SIM) or selected reaction monitoring (SRM) modes for targeted analysis. In contrast, HRMS allows qualitative and quantitative analysis for target, non-target and unknown compounds [84-86].

LC-HRMS, with electrospray ionization sources working in the positive mode, has been used for identification and quantitation for both single fumonisins analysis and also for fumonisins included in multi-mycotoxin analysis. In all cases the pseudomolecular $[M+H]^+$ ions are monitored.

The use of LC-HRMS methods for the analysis of single or a small group of fumonisins has been mainly focused to their characterization in food matrices or to obtaining reference standards. As an example, Table 3 summarizes a selection of LC-HRMS methods for the characterization and analysis of fumonisins and/or multi-mycotoxin analysis [87-91].

Fumonisin B_1 (FB_1) and fumonisin B_5 (FB_5) along with an important number of isomers produced in a solid rice culture infected with *Fusarium verticillioides* have been investigated by using LC-HRMS [87, 92]. Separation of FB_1 and FB_5 isomers was carried out on a YMC-Pack J'sphere ODS H80 narrow-bore column with a flat gradient. Exact mass measurements obtained by LC-TOFMS (Agilent 6210) of the protonated molecules led to the identification of signals. Fragmentation patterns from ion trap MS/MS spectra were used to confirm the identity of the compounds. Also, LC-HRMS with an Orbitrap (Thermo Fisher Scientific), a heated electrospray ion source and a reversed-phase column (Gemini C18; Phenomenex) was used to isolate and characterize partially hydrolyzed fumonisins ($PHFB_1$ and $PHFB_2$) and hydrolyzed fumonisins (HFB_1 and HFB_2) obtained by chemical hydrolysis of pure fumonisins (FB_1 and FB_2) to be used as reference standards [90]. The method was also applied to the simultaneous determination of fumonisins and their partially and totally hydrolyzed derivatives in naturally contaminated samples of maize and maize-based products. As a single stage Orbitrap mass analyzer was employed, in order to get an LC–HRMS detection method in compliance with the European official guidelines for confirmatory analysis [80], i.e. the use of at least two diagnostic ions and measurement of both ions with resolution >10 000 FWHM (full width at half maximum) and mass accuracy <5 ppm, a high energy collision dissociation (HCD) cell was used.

Fumonisins A-series (FAs) in a reference material of naturally contaminated corn was also characterized using LC-HRMS. In previous reports, several fumonisin analogs were detected by quadrupole tandem mass spectrometry (MS/MS). The nominal mass obtained by this low resolution instrument has poor predictive power for unknown compounds due to the potentially numerous candidates with similar chemical formulae. Consequently, there was little evidence that the compounds detected in such reports were fumonisin analogs. The use of a Q-Exactive (Thermo Fisher Scientific), which is an Orbitrap MS equipped with a quadrupole mass filter and a collision cell, along with the use of synthesized reference standards, gave enough evidence to confirm that the compounds were FA_1, FA_2 and FA_3. Besides these FAs, FB_1, FB_2 and FB_3 were detected in the corn sample [93, 94].

Currently, there is an increasing trend to include fumonisins in multi-mycotoxin assays in order to get simple, broad scope procedures which allow accurate determination of major fungal toxins even at low levels in crude extracts, i.e. without labor/cost demanding clean-up step. An important number of the published multi-mycotoxin methods use tandem mass spectrometry (triple quadrupole or Q-trap mass analyzers are most common) for selective detection and confirmation of target mycotoxins, as previously commented. However under these conditions, retrospective data analysis of samples is not possible and only target analytes can be detected. Some research groups have evaluated HRMS to obtain not only full spectral information but also for the detection of unknown signals in sample extracts on the basis of accurate mass measurement, and eventually, fragments induced in the ion source [83, 95]. Recently, some multi-mycotoxin LC-HRMS methods including fumonisins have been developed [86, 88, 89, 91, 95-97].

One of the most important challenges in developing multi-mycotoxin analytical methods is related with the wide range of different physicochemical properties in terms of pH stability, solubility, diversity of chemical structure and molecular weight (small/large molecules) that mycotoxins exhibit. The contamination of biological samples by fungal toxins is often heterogeneous (e.g. trichothecenes, aflatoxins, ochratoxins, ergot alkaloids, enniatins, fumonisins, etc.) and consequently makes further analyses for detection and quantification more complicated [96, 97]. Particularly, fumonisins are pH-sensitive ionic compounds, and they need analytical conditions that are not favorable for other mycotoxins. For this reason, some authors preferred as a compromise two chromatographic runs, each providing optimal detection conditions for target analytes [95].

Both TOF and Orbitrap mass spectrometers have been evaluated for multi-mycotoxin analytical methods. Zacharisova et al., [95] compared the performance of a Waters LCT Premier XE TOFMS and an Exactive Orbitrap mass spectrometer from Thermo Fisher Scientific. LC-TOF-HRMS working at its maximum resolving power (12,000 FWHM), showed high chemical noise and strong analytes signal suppression due to co-eluting matrix. So, with this instrument direct analysis of crude extract was not possible and a cleanup procedure had to be introduced. In-source fragmentation was tried, in order to obtain other diagnostic ions in addition to pseudomolecular ion, but fragments of mycotoxins in presence of the matrix were possible to identify with the adequate mass accuracy only at fairly high concentration levels. Better results were obtained using LC-HRMS with the Orbitrap mass spectrometer. Working at a resolving power of 100,000 FWHM, no matrix effect was observed and crude extracts could be analyzed directly, making the procedure simpler. Under these conditions, good mass accuracy was obtained, and for instance for FB_1 was in the range -0.7 up to +0.3 ppm.

When working with high resolving powers it is necessary to ensure that sufficient data points (12–15) are obtained. Column dimension is a critical parameter that should be evaluated when working HRMS. At a higher resolution, relatively slow scan speeds should be set and broad enough chromatographic peaks are thus needed. At a resolution power of 100,000 FWHM, a column with 50 x 4.6 mm and 5 μm particle size was found to give the best results [95, 97].

HRMS coupled to LC has become an invaluable tool for the analysis of metabolites in pharmaceutical, food and life sciences. The amount of data generated from this type of analysis is on a continuous rise due to the increasing number of analyses conducted on one hand, and an increasing mass resolution of the mass spectrometer on the other hand. As a consequence, the use of tools that are capable of automatically processing these large numbers of LC-HRMS datasets along with well-planned work-flows is becoming even more necessary [98]. Most multi-mycotoxin analytical methods include target screening approaches that combine full scan acquisition with in-source fragmentation or full scan acquisition and data dependent MS^n analysis. In this last mode, the acquisition software probes the MS spectra in real time on a full scan and allows accurate mass measurement. With the accurate mass the instrument is capable of finding true unknowns since the method does not require any pre-selection of masses. The instrument is initially set to operate in full scan mode until a parent ion appears to preset the instrument, which switches into the product-ion mode (MS^n). Using this strategy, coeluting matrix compounds from the matrix or noisy peaks can be easily excluded, facilitating the identification and quantification. Library search of the spectra using spectral databases obtained can lead to the identification of unknowns [86, 88, 91, 95, 97].

An important issue for LC-HRMS methods is the compliance with requirements for performance characteristics stated in Commission Decision 2002/657/EC [80]. Besides with good repeatability of retention times (RDS < 2.5%), for mycotoxins three identification points (IPs) are required. The higher is the resolving power of the mass spectrometer, the lower number of diagnostic ions is needed to meet this criterion. According to the 2002 Commission Decision mentioned above, in HRMS, the resolution shall typically be greater than 10,000 for the entire mass range at 10% valley. For many common TOF mass analyzers, this value, which roughly equals to 20,000 when expressed as FWHM, is not achieved. Triple quadrupole mass spectrometers, can easily reach for IPs working on selected reaction monitoring mode, thereby this instrument fulfill the criterion established.

However, full scan spectra obtained by Orbitrap did not permit to obtain enough IPs, although the instrument worked at ultra high mass resolving power settings of 100,000 FWHM and the mass error exceeding slightly 1 ppm. Obviously, when the analysis is done by combining full scan and data dependent analysis (DDA), the number of IPs can be increased. It should be noted, that in 2002/657/EC no criterions have been set for mass accuracy, which is typically less than 5 ppm for such instruments. So, the need to update the identification criteria is nowadays widely discussed by the scientific community [86, 90, 95, 97, 99].

CONCLUSION

As reviewed, fumonisins, mycotoxins produced by *Fusarium Fungi*, are commonly found in maize and many other foodstuffs. Because of the available evidence of their toxicity, maximum residue levels have been established by several countries. In the European Union, these levels have been decreed in maize and its derived products. So today, the development of analytical methodologies for the routine control of fumonisins, among other mycotoxins, in feed and foodstuffs is necessary.

Among the several methodologies than can be employed for the analysis of fumonisins, liquid chromatography coupled to either mass spectrometry (LC-MS) or high resolution mass spectrometry (LC-HRMS) are the techniques of choice for the analysis and characterization of fumonisins in foodstuffs. The state-of-the-art of LC-MS and LC-HRMS for the analysis and characterization of fumonisins in food-based products and other matrices has been addressed in this chapter. LC-MS chromatographic conditions, ionization sources, and MS and HRMS analyzers frequently used, as well as strategies for the structural characterization and the qualitative and quantitative analysis of fumonisins, have been discussed by means of relevant applications. LC-MS/MS working in SRM acquisition mode is the method of choice when dealing with multi-mycotoxin analysis mainly using triple quadrupole mass analyzers, although other ion-trap based spectrometers are also employed. Regarding HRMS, Orbitrap mass analyzers are becoming very popular for the LC-HRMS characterization of fumonisins and mycotoxins in general because of their higher resolution power and accurate mass measurements.

However, despite the significant advances in chromatographic separations and MS techniques, sample treatment is still one of the most significant parts of the entire analytical process. Considerable efforts have therefore been made in the past years to simplify this step and to develop fast, accurate and precise methodologies able to allow the analytical determination of target mycotoxins without compromising the integrity of the extraction process. Several extraction techniques have been addressed in this chapter for the clean-up and extraction of fumonisins including simple solid-liquid extraction (SLE), solid-phase extraction (SPE), immunoaffinity procedures, the use of molecularly imprinted polymers (MIPs), and QuEChERS methods.

ACKNOWLEDGMENTS

This work has been funded by the Spanish Ministry of Economy and Competitiveness under the project CTQ2012-30836, and from the Agency for Administration of University and Research Grants (Generalitat de Catalunya, Spain) under the project 2014 SGR-539.

REFERENCES

[1] Pitt, J. I., Basilico, J. C., Abarca, M. L. & Lopez, C. (2000). Mycotoxins and toxigenic fungi. *Med. Mycol.*, 38, 41-46.

[2] Pitt, J. I. (2000). Toxigenic fungi and mycotoxins. *Br. Med. Bull.*, 56, 184-192.

[3] Garcia-Cela, E., Ramos, A. J., Sanchis, V. & Marin, S. (2012). Emerging risk management metrics in food safety: FSO, PO. How do they apply to the mycotoxin hazard? *Food Control*, 25, 797-808.

[4] Gelderblom, W. C. A., Jaskiewicz, K., Marasas, W. F. O., Thiel, P. G., Horak, R. M., Vleggaar, R. & Kriek, N. P. J. (1988). Fumonisins-novel mycotoxins with cancer-promoting activity produced by Fusarium moniliforme. *Appl. Environ. Microbiol.*, 54, 1806-1811.

[5] Gelderblom, W. C. A., Marasas, W. F. O., Vleggaar, R., Thiel, P. G. & Cawood, M. E. (1992). Fumonisins: isolation, chemical characterization and biological effects. *Mycopathologia*, 117, 11-16.

[6] Soriano, J. M. & Dragacci, S. (2004). Occurrence of fumonisins in foods. *Food Res. Int.*, 37, 985-1000.

[7] WHO-IPCS (World Health Organization e International Programme on Chemical Safety), Fumonisin B1. Environmental health criteria 219, World Health Organization, Geneva, Switzerland 2000.

[8]   European Commission, Opinion of the Scientific Committee on Food on *Fusarium* toxins - Part 3: Fumonisin B1 (FB1), European Commission, Brussels, Belgium 2000.

[9]   Norhasima, W. M. W., Abdulamir, A. S., abu Bakar, F., Son, R. & Norhafniza, A. (2009). The health and toxic adverse effects of Fusarium fungal mycotoxin, fumonisins, on human population. *Am. J. Infect. Dis.*, 5, 273-281.

[10]   Heperkan, D., Gueler, F. K. & Oktay, H. I. (2012). Mycoflora and natural occurrence of aflatoxin, cyclopiazonic acid, fumonisin and ochratoxin A in dried figs. *Food Addit. Contam. , Part A*, 29, 277-286.

[11]   Haas, D., Pfeifer, B., Reiterich, C., Partenheimer, R., Reck, B. & Buzina, W. (2013). Identification and quantification of fungi and mycotoxins from Pu-erh tea. *Int. J. Food Microbiol.*, 166, 316-322.

[12]   Kaya, S. B. & Tosun, H. (2013). Occurrence of total aflatoxin, ochratoxin A and fumonisin in some organic foods. *J. Pure Appl. Microbiol.*, 7, 2925-2932.

[13]   Ashiq, S., Hussain, M. & Ahmad, B. (2014). Natural occurrence of mycotoxins in medicinal plants: A review. *Fungal Genet. Biol.*, 66, 1-10.

[14]   Commission Regulation (EC) No 1126/2007 of 28 September 2007 amending Regulation (EC) No 1881/2006 setting maximum levels for certain contaminants in foodstuffs as regards *Fusarium* toxins in maize and maize products (2007) *Off. J. Eur. Union,* L255, 14.

[15]   Food and Drug Administration (FDA). Guidance for Industry: Fumonisin levels in human foods and animal feeds; Final guidance. Food and Drug Administration Center for Food Safety and Applied Nutrition Center for Veterinary Medicine (2001). Available at: http://www.fda.gov/Food/GuidanceRegulation/GuidanceDocumentsReg ulatoryInformation/ChemicalContaminantsMetalsNaturalToxinsPesticid es/ucm109231.htm (accessed 1st January 2015).

[16]   Cano-Sancho, G., Ramos, A. J., Marin, S. & Sanchis, V. (2012). Occurrence of fumonisins in Catalonia (Spain) and an exposure assessment of specific population groups. *Food Addit. Contam. , Part A*, 29, 799-808.

[17]   Muscarella, M., Lo Magro, S., Nardiello, D., Palermo, C. & Centonze, D. (2011). Determination of fumonisins B1 and B2 in maize food products by a new analytical method based on high-performance liquid chromatography and fluorimetric detection with post-column derivatization. *Methods Mol. Biol. (N. Y. , NY, U. S.)*, 739, 187-194.

[18] Ndube, N., van der Westhuizen, L., Green, I. R. & Shephard, G. S. (2011). HPLC determination of fumonisin mycotoxins in maize: A comparative study of naphthalene-2,3-dicarboxaldehyde and o-phthaldialdehyde derivatization reagents for fluorescence and diode array detection. *J. Chromatogr. B: Anal. Technol. Biomed. Life Sci.*, 879, 2239-2243.

[19] de Girolamo, A., Pereboom-de Fauw, D., Sizoo, E., van Egmond, H. P., Gambacorta, L., Bouten, K., Stroka, J., Visconti, A. & Solfrizzo, M. (2010). Determination of fumonisins B1 and B2 in maize-based baby food products by HPLC with fluorimetric detection after immunoaffinity column clean-up. *World Mycotoxin J.*, 3, 135-146.

[20] Silva, L., Fernandez-Franzon, M., Font, G., Pena, A., Silveira, I., Lino, C. & Manes, J. (2008). Analysis of fumonisins in corn-based food by liquid chromatography with fluorescence and mass spectrometry detectors. *Food Chem.*, 112, 1031-1037.

[21] Paepens, C., De Saeger, S., Van Poucke, C., Dumoulin, F., Van Calenbergh, S. & Van Peteghem, C. (2005). Development of a liquid chromatography/tandem mass spectrometry method for the quantification of fumonisin B1, B2 and B3 in cornflakes. *Rapid Commun. Mass Spectrom.*, 19, 2021-2029.

[22] Paepens, C., De Saeger, S., Sibanda, L., Barna-Vetro, I., Anselme, M., Larondelle, Y. & Van Peteghem, C. (2005). Evaluation of Fumonisin Contamination in Cornflakes on the Belgian Market by "Flow-Through" Assay Screening and LC-MS/MS Analyses. *J. Agric. Food Chem.*, 53, 7337-7343.

[23] Faberi, A., Foglia, P., Pastorini, E., Samperi, R. & Lagana, A. (2005). Determination of type B fumonisin mycotoxins in maize and maize-based products by liquid chromatography/tandem mass spectrometry using a QqQlinear ion trap mass spectrometer. *Rapid Commun. Mass Spectrom.*, 19, 275-282.

[24] Zitomer, N. C., Glenn, A. E., Bacon, C. W. & Riley, R. T. (2008). A single extraction method for the analysis by liquid chromatography/tandem mass spectrometry of fumonisins and biomarkers of disrupted sphingolipid metabolism in tissues of maize seedlings. *Anal. Bioanal. Chem.*, 391, 2257-2263.

[25] Bryla, M., Jedrzejczak, R., Roszko, M., Szymczyk, K., Obiedzinski, M. W., Sekul, J. & Rzepkowska, M. (2013). Application of molecularly imprinted polymers to determine B1, B2, and B3 fumonisins in cereal products. *J. Sep. Sci.*, 36, 578-584.

[26] D'Arco, G., Fernandez-Franzon, M., Font, G., Damiani, P. & Manes, J. (2008). Analysis of fumonisins B1, B2 and B3 in corn-based baby food by pressurized liquid extraction and liquid chromatography/tandem mass spectrometry. *J. Chromatogr. A*, 1209, 188-194.

[27] Selim, M. I., El Sharkawy, S. H. & Popendorf, W. J. (1996). Supercritical Fluid Extraction of Fumonisin B1 from Grain Dust. *J. Agric. Food Chem.*, 44, 3224-3229.

[28] Seefelder, W., Hartl, M. & Humpf, H. U. (2001). Determination of N-(Carboxymethyl)fumonisin B1 in Corn Products by Liquid Chromatography/Electrospray Ionization Mass Spectrometry. *J. Agric. Food Chem.*, 49, 2146-2151.

[29] Khayoon, W. S., Saad, B., Salleh, B., Ismail, N. A., Abdul Manaf, N. H. & Latiff, A. A. (2010). A reversed phase high performance liquid chromatography method for the determination of fumonisins B1 and B2 in food and feed using monolithic column and positive confirmation by liquid chromatography/tandem mass spectrometry. *Anal. Chim. Acta*, 679, 91-97.

[30] Gazzotti, T., Zironi, E., Lugoboni, B., Barbarossa, A., Piva, A. & Pagliuca, G. (2011). Analysis of fumonisins B1, B2 and their hydrolysed metabolites in pig liver by LC-MS/MS. *Food Chem.*, 125, 1379-1384.

[31] Royer, D., Humpf, H. U. & Guy, P. A. (2004). Quantitative analysis of Fusarium mycotoxins in maize using accelerated solvent extraction before liquid chromatography/atmospheric pressure chemical ionization tandem mass spectrometry. *Food Addit. Contam.*, 21, 678-692.

[32] Seefelder, W., Gossmann, M. & Humpf, H. U. (2002). Analysis of Fumonisin B1 in Fusarium proliferatum-Infected Asparagus Spears and Garlic Bulbs from Germany by Liquid Chromatography-Electrospray Ionization Mass Spectrometry. *J. Agric. Food Chem.*, 50, 2778-2781.

[33] De Smet, D., Dubruel, P., Van Peteghem, C., Schacht, E. & De Saeger, S. (2009). Molecularly imprinted solid-phase extraction of fumonisin B analogues in bell pepper, rice and corn flakes. *Food Addit. Contam. , Part A*, 26, 874-884.

[34] Silva, L., Fernandez-Franzon, M., Font, G., Pena, A., Silveira, I., Lino, C. & Manes, J. (2008). Analysis of fumonisins in corn-based food by liquid chromatography with fluorescence and mass spectrometry detectors. *Food Chem.*, 112, 1031-1037.

[35] Gazzotti, T., Lugoboni, B., Zironi, E., Barbarossa, A., Serraino, A. & Pagliuca, G. (2009). Determination of fumonisin B1 in bovine milk by LC-MS/MS. *Food Control*, 20, 1171-1174.

[36] Silva, L. J. G., Pena, A., Lino, C. M., Fernandez, M. F. & Manes, J. (2010). Fumonisins determination in urine by LC-MS-MS. *Anal. Bioanal. Chem.*, 396, 809-816.

[37] Kong, W., Xie, T., Li, J., Wei, J., Qiu, F., Qi, A., Zheng, Y. & Yang, M. (2012). Analysis of fumonisins B1 and B2 in spices and aromatic and medicinal herbs by HPLC-FLD with on-line post-column derivatization and positive confirmation by LC-MS/MS. *Analyst (Cambridge, U. K.)*, 137, 3166-3174.

[38] Desmarchelier, A., Oberson, J. M., Tella, P., Gremaud, E., Seefelder, W. & Mottier, P. (2010). Development and Comparison of Two Multiresidue Methods for the Analysis of 17 Mycotoxins in Cereals by Liquid Chromatography Electrospray Ionization Tandem Mass Spectrometry. *J. Agric. Food Chem.*, 58, 7510-7519.

[39] Tamura, M., Uyama, A. & Mochizuki, N. (2011). Development of a multi-mycotoxin analysis in beer-based drinks by a modified QuEChERS method and ultra-high-performance liquid chromatography coupled with tandem mass spectrometry. *Anal. Sci.*, 27, 629-635.

[40] Yogendrarajah, P., Van Poucke, C., De Meulenaer, B. & De Saeger, S. (2013). Development and validation of a QuEChERS based liquid chromatography tandem mass spectrometry method for the determination of multiple mycotoxins in spices. *J. Chromatogr. A*, 1297, 1-11.

[41] Arroyo-Manzanares, N., Garcia-Campana, A. M. & Gamiz-Gracia, L. (2013). Multiclass mycotoxin analysis in Silybum marianum by ultra high performance liquid chromatography-tandem mass spectrometry using a procedure based on QuEChERS and dispersive liquid-liquid microextraction. *J. Chromatogr. A*, 1282, 11-19.

[42] Bolechova, M., Benesova, K., Belakova, S., Caslavsky, J., Pospichalova, M. & Mikulikova, R. (2015). Determination of seventeen mycotoxins in barley and malt in the Czech Republic. *Food Control*, 47, 108-113.

[43] Arroyo-Manzanares, N., Huertas-Perez, J. F., Garcia-Campana, A. M. & Gamiz-Gracia, L. (2014). Simple methodology for the determination of mycotoxins in pseudocereals, spelt and rice. *Food Control*, 36, 94-101.

[44] Desmarchelier, A., Tessiot, S., Bessaire, T., Racault, L., Fiorese, E., Urbani, A., Chan, W. C., Cheng, P. & Mottier, P. (2014). Combining the quick, easy, cheap, effective, rugged and safe approach and clean-up by immunoaffinity column for the analysis of 15 mycotoxins by isotope dilution liquid chromatography tandem mass spectrometry. *J. Chromatogr. A*, 1337, 75-84.

[45] Vekiru, E., Fuchs, E., Schatzmayr, G., Taeubel, M., Binder, E. M. & Krska, R. (2003). Determination of fumonisins and hydrolyzed fumonisin B1 in microbial culture media by LC/ESI-MS. *Mycotoxin Res.*, 19, 198-202.

[46] Ren, Y., Zhang, Y., Lai, S., Han, Z. & Wu, Y. (2011). Simultaneous determination of fumonisins B1, B2 and B3 contaminants in maize by ultra high-performance liquid chromatography tandem mass spectrometry. *Anal. Chim. Acta*, 692, 138-145.

[47] Li, C., Wu, Y. L., Yang, T. & Huang-Fu, W. G. (2012). Rapid Determination of Fumonisins B1 and B2 in Corn by Liquid Chromatography-Tandem Mass Spectrometry with Ultrasonic Extraction. *J. Chromatogr. Sci.*, 50, 57-63.

[48] Bergmann, D., Huebner, F. & Humpf, H. U. (2013). Stable Isotope Dilution Analysis of Small Molecules with Carboxylic Acid Functions Using 18O Labeling for HPLC-ESI-MS/MS: Analysis of Fumonisin B1. *J. Agric. Food Chem.*, 61, 7904-7908.

[49] Bartok, T., Szecsi, A., Szekeres, A., Mesterhazy, A. & Bartok, M. (2006). Detection of new fumonisin mycotoxins and fumonisin-like compounds by reversed-phase high-performance liquid chromatography/electrospray ionization ion trap mass spectrometry. *Rapid Commun. Mass Spectrom.*, 20, 2447-2462.

[50] Sorensen, L. K. & Elbaek, T. H. (2005). Determination of mycotoxins in bovine milk by liquid chromatography tandem mass spectrometry. *J. Chromatogr. B: Anal. Technol. Biomed. Life Sci.*, 820, 183-196.

[51] Cavaliere, C., Foglia, P., Pastorini, E., Samperi, R. & Lagana, A. (2005). Development of a multiresidue method for analysis of major Fusarium mycotoxins in corn meal using liquid chromatography/tandem mass spectrometry. *Rapid Commun. Mass Spectrom.*, 19, 2085-2093.

[52] Lattanzio, V. M. T., Solfrizzo, M., Powers, S. & Visconti, A. (2007). Simultaneous determination of aflatoxins, ochratoxin A and Fusarium toxins in maize by liquid chromatography/tandem mass spectrometry after multitoxin immunoaffinity cleanup. *Rapid Commun. Mass Spectrom.*, 21, 3253-3261.

[53] Monbaliu, S., Van Poucke, C., Van Peteghem, C., Van Poucke, K., Heungens, K. & De Saeger, S. (2009). Development of a multi-mycotoxin liquid chromatography/tandem mass spectrometry method for sweet pepper analysis. *Rapid Commun. Mass Spectrom.*, 23, 3-11.

[54] Sorensen, L. M., Mogensen, J. & Nielsen, K. F. (2010). Simultaneous determination of ochratoxin A, mycophenolic acid and fumonisin B2 in meat products. *Anal. Bioanal. Chem.*, 398, 1535-1542.

[55] Devreese, M., De Baere, S., De Backer, P. & Croubels, S. (2012). Quantitative determination of several toxicological important mycotoxins in pig plasma using multi-mycotoxin and analyte-specific high performance liquid chromatography-tandem mass spectrometric methods. *J. Chromatogr. A*, 1257, 74-80.

[56] Oueslati, S., Romero-Gonzalez, R., Lasram, S., Frenich, A. G. & Vidal, J. L. M. (2012). Multi-mycotoxin determination in cereals and derived products marketed in Tunisia using ultra-high performance liquid chromatography coupled to triple quadrupole mass spectrometry. *Food Chem. Toxicol.*, 50, 2376-2381.

[57] Tamura, M., Takahashi, A., Uyama, A. & Mochizuki, N. (2012). A method for multiple mycotoxin analysis in wines by solid phase extraction and multifunctional cartridge purification, and ultra-high-performance liquid chromatography coupled to tandem mass spectrometry. *Toxins*, 4, 476-486.

[58] Soleimany, F., Jinap, S. & Abas, F. (2012). Determination of mycotoxins in cereals by liquid chromatography tandem mass spectrometry. *Food Chem.*, 130, 1055-1060.

[59] Wang, Y., Xiao, C., Guo, J., Yuan, Y., Wang, J., Liu, L. & Yue, T. (2013). Development and Application of a Method for the Analysis of 9 Mycotoxins in Maize by HPLC-MS/MS. *J. Food Sci.*, 78, M1752-M1756.

[60] Song, S., Ediage, E. N., Wu, A. & De Saeger, S. (2013). Development and application of salting-out assisted liquid/liquid extraction for multi-mycotoxin biomarkers analysis in pig urine with high performance liquid chromatography/tandem mass spectrometry. *J. Chromatogr. A*, 1292, 111-120.

[61] Vaclavikova, M., MacMahon, S., Zhang, K. & Begley, T. H. (2013). Application of single immunoaffinity clean-up for simultaneous determination of regulated mycotoxins in cereals and nuts. *Talanta*, 117, 345-351.

[62] Liao, C. D., Wong, J. W., Zhang, K., Hayward, D. G., Lee, N. S. & Trucksess, M. W. (2013). Multi-mycotoxin analysis of finished grain and nut products using high-performance liquid chromatography-triple-quadrupole mass spectrometry. *J. Agric. Food Chem.*, 61, 4771-4782.

[63] Abia, W. A., Warth, B., Sulyok, M., Krska, R., Tchana, A. N., Njobeh, P. B., Dutton, M. F. & Moundipa, P. F. (2013). Determination of multi-mycotoxin occurrence in cereals, nuts and their products in Cameroon by liquid chromatography tandem mass spectrometry (LC-MS/MS). *Food Control*, 31, 438-453.

[64] Zhang, K., Wong, J. W., Krynitsky, A. J. & Trucksess, M. W. (2014). Determining Mycotoxins in Baby Foods and Animal Feeds Using Stable Isotope Dilution and Liquid Chromatography Tandem Mass Spectrometry. *J. Agric. Food Chem.*, 62, 8935-8943.

[65] Pizzutti, I. R., de Kok, A., Scholten, J., Righi, L. W., Cardoso, C. D., Necchi Rohers, G. & da Silva, R. C. (2014). Development, optimization and validation of a multimethod for the determination of 36 mycotoxins in wines by liquid chromatography-tandem mass spectrometry. *Talanta*, 129, 352-363.

[66] Skrbic, B., Zivancev, J. & Godula, M. (2014). Multimycotoxin analysis of crude extracts of nuts with ultra-high performance liquid chromatography/tandem mass spectrometry. *J. Food Compos. Anal.*, 34, 171-177.

[67] Blajet-Kosicka, A., Kosicki, R., Twaruzek, M. & Grajewski, J. (2014). Determination of moulds and mycotoxins in dry dog and cat food using liquid chromatography with mass spectrometry and fluorescence detection. *Food Addit. Contam. , Part B*, Ahead-

[68] Yibadatihan, S., Jinap, S. & Mahyudin, N. A. (2014). Simultaneous determination of multi-mycotoxins in palm kernel cake (PKC) using liquid chromatography-tandem mass spectrometry (LC-MS/MS). *Food Addit. Contam. , Part A*, Ahead-

[69] Williams, L. D., Glenn, A. E., Zimeri, A. M., Bacon, C. W., Smith, M. A. & Riley, R. T. (2007). Fumonisin Disruption of Ceramide Biosynthesis in Maize Roots and the Effects on Plant Development and Fusarium verticillioides-Induced Seedling Disease. *J. Agric. Food Chem.*, 55, 2937-2946.

[70] Riley, R. T., Enongene, E., Voss, K. A., Norred, W. P., Meredith, F. I., Sharma, R. P., Spitsbergen, J., Williams, D. E., Carlson, D. B. & Merrill, A. H., Jr. (2001). Sphingolipid perturbations as mechanisms for fumonisin carcinogenesis. *Environ. Health Perspect. Suppl.*, 109, 301-308.

[71] Glenn, A. E., Zitomer, N. C., Zimeri, A. M., Williams, L. D., Riley, R. T. & Proctor, R. H. (2008). Transformation-mediated complementation of a FUM gene cluster deletion in Fusarium verticillioides restores both

fumonisin production and pathogenicity on maize seedlings. *Mol. Plant-Microbe Interact.*, 21, 87-97.

[72] Núñez, O., Gallart-Ayala, H., Martins, C. P. B. & Lucci, P. (2012). New trends in fast liquid chromatography for food and environmental analysis. *J. Chromatogr. A*, 1228, 298-323.

[73] Núñez, O., Gallart-Ayala, H., Martins, C.P.B. & Lucci, P. (Editors) (2015), Fast Liquid Chromatography-Mass Spectrometry Methods in Food and Environmental Analysis, Imperial College Press, London, UK. (ISBN 978-1-78326-493-3)

[74] D'Orazio, G., Rocco, A. & Fanali, S. (2012). Fast-liquid chromatography using columns of different internal diameters packed with sub-2 Î¼m silica particles. *J. Chromatogr. A*, 1228, 213-220.

[75] Schappler, J., Veuthey, J-L. & Guillarme, D. (2015) UHPLC Separations Using Sub-2μm Particle Size Columns, in: Núñez O., Gallart-Ayala, H., Martins, C.P.B. & Lucci, P. (Editors), Fast Liquid Chromatography-Mass Spectrometry Methods in Food and Environmental Analysis, Imperial College Press, London, UK, pp. 3-32.

[76] Salisbury, J. J. (2008). Fused-core particles: A practical alternative to sub-2 micron particles. *J. Chromatogr. Sci.*, 46, 883-886.

[77] Fekete, S., Fekete, J. & Ganzler, K. (2009). Shell and small particles; Evaluation of new column technology. *J. Pharm. Biomed. Anal.*, 49, 64-71.

[78] Núñez, O. & Gallart-Ayala, H. (2015) Core-shell Column Technology in Fast Liquid Chromatography, in: Núñez O., Gallart-Ayala, H., Martins, C.P.B. & Lucci, P. (Editors), Fast Liquid Chromatography-Mass Spectrometry Methods in Food and Environmental Analysis, Imperial College Press, London, UK, pp. 33-56.

[79] Núñez, O., Checa, A. & Gallart-Ayala, H. (2013) Fluorinated Stationary Phases on Liquid Chromatography: Preparation, Properties and Applications, in Ramos, F. (Editor), Liquid Chromatography: Principles, Technology and Applications, Nova Science Publishers, Inc, Hauppauge, NY, USA, pp. 37-55. (ISBN 978-1-62618-739-9)

[80] Commission Decicion 2002/657/EC of 12 of August 2002 implementing Council Directive 96/23/EC concerning the performance of analytical methods and the interpretation of results, Off. J. Eur. Commun. L 221 (2002) 8.

[81] Han, Z., Ren, Y., Liu, X., Luan, L. & Wu, Y. (2010). A reliable isotope dilution method for simultaneous determination of fumonisins B1, B2 and B3 in traditional Chinese medicines by ultra-high-performance

liquid chromatography-tandem mass spectrometry. *J. Sep. Sci.*, 33, 2723-2733.

[82] Dall'Asta, C., Mangia, M., Berthiller, F., Molinelli, A., Sulyok, M., Schuhmacher, R., Krska, R., Galaverna, G., Dossena, A. & Marchelli, R. (2009). Difficulties in fumonisin determination: the issue of hidden fumonisins. *Anal. Bioanal. Chem.*, 395, 1335-1345.

[83] Berthiller, F., Brera, C., Crews, C., Iha, M. H., Krska, R., Lattanzio, V. M. T., MacDonald, S., Malone, R. J., Maragos, C., Solfrizzo, M., Stroka, J. & Whitaker, T. B. (2015). Developments in mycotoxin analysis: an update for 2013-2014. *World Mycotoxin J.*, 8, 5-36.

[84] Meyer, M. R., Helfer, A. G. & Maurer, H. H. (2014). Current position of high-resolution MS for drug quantification in clinical & forensic toxicology. *Bioanalysis*, 6, 2275-2284.

[85] Wu, A. H., Gerona, R., Armenian, P., French, D., Petrie, M. & Lynch, K. L. (2012). Role of liquid chromatography-high-resolution mass spectrometry (LC-HR/MS) in clinical toxicology. *Clin. Toxicol.*, 50, 733-742.

[86] Rubert, J., Soler, C., Marin, R., James, K. J. & Manes, J. (2013). Mass spectrometry strategies for mycotoxins analysis in European beers. *Food Control*, 30, 122-128.

[87] Bartok, T., Tolgyesi, L., Szekeres, A., Varga, M., Bartha, R., Szecsi, A., Bartok, M. & Mesterhazy, A. (2010). Detection and characterization of twenty-eight isomers of fumonisin B1 (FB1) mycotoxin in a solid rice culture infected with Fusarium verticillioides by reversed-phase high-performance liquid chromatography/electrospray ionization time-of-flight and ion trap mass spectrometry. *Rapid Commun. Mass Spectrom.*, 24, 35-42.

[88] Rubert, J., Manes, J., James, K. J. & Soler, C. (2011). Application of hybrid linear ion trap-high resolution mass spectrometry to the analysis of mycotoxins in beer. *Food Addit. Contam. , Part A*, 28, 1438-1446.

[89] Perez-Ortega, P., Gilbert-Lopez, B., Garcia-Reyes, J. F., Ramos-Martos, N. & Molina-Diaz, A. (2012). Generic sample treatment method for simultaneous determination of multiclass pesticides and mycotoxins in wines by liquid chromatography-mass spectrometry. *J. Chromatogr. A*, 1249, 32-40.

[90] De Girolamo, A., Lattanzio, V. M. T., Schena, R., Visconti, A. & Pascale, M. (2014). Use of liquid chromatography-high-resolution mass spectrometry for isolation and characterization of hydrolyzed fumonisins

and relevant analysis in maize-based products. *J. Mass Spectrom.*, 49, 297-305.

[91] Jia, W., Chu, X., Ling, Y., Huang, J. & Chang, J. (2014). Multi-mycotoxin analysis in dairy products by liquid chromatography coupled to quadrupole orbitrap mass spectrometry. *J. Chromatogr. A*, 1345, 107-114.

[92] Bartok, T., Tolgyesi, L., Szecsi, A., Varga, J., Bartok, M., Mesterhazy, A., Gyimes, E. & Veha, A. (2013). Identification of unkown isomers of fumonisin B5 mycotoxin in a *Fusarium Verticillioides* culture by highperformance liquid chromatography/electrospray ionization time-of-flight and ion trap mass spectrometry. *J. Liq. Chromatogr. Relat. Technol.*, 36, 1549-1561.

[93] Tamura, M., Mochizuki, N., Nagatomi, Y., Toriba, A. & Hayakawa, K. (2014). Characterization of fumonisin A-Series by high-resolution liquid chromatography-orbitrap mass spectrometry. *Toxins*, 6, 2580-2593.

[94] Tamura, M., Mochizuki, N., Nagatomi, Y., Harayama, K., Toriba, A. & Hayakawa, K. (2015). Identification and quantification of fumonisin A1, A2, and A3 in corn by high-resolution liquid chromatography-Orbitrap mass spectrometry. *Toxins*, 7, 582-592.

[95] Zachariasova, M., Lacina, O., Malachova, A., Kostelanska, M., Poustka, J., Godula, M. & Hajslova, J. (2010). Novel approaches in analysis of Fusarium mycotoxins in cereals employing ultra performance liquid chromatography coupled with high resolution mass spectrometry. *Anal. Chim. Acta*, 662, 51-61.

[96] Herebian, D., Zuehlke, S., Lamshoeft, M. & Spiteller, M. (2009). Multi-mycotoxin analysis in complex biological matrices using LC-ESI/MS: experimental study using triple stage quadrupole and LTQ-orbitrap. *J. Sep. Sci.*, 32, 939-948.

[97] Ates, E., Mittendorf, K., Stroka, J. & Senyuva, H. (2013). Determination of fusarium mycotoxins in wheat, maize and animal feed using on-line clean-up with high resolution mass spectrometry. *Food Addit. Contam. , Part A*, 30, 156-165.

[98] Krishnan, S., Verheij, E. E. R., Bas, R. C., Hendriks, M. W. B., Hankemeier, T., Thissen, U. & Coulier, L. (2013). Pre-processing liquid chromatography/high-resolution mass spectrometry data: extracting pure mass spectra by deconvolution from the invariance of isotopic distribution. *Rapid Commun. Mass Spectrom.*, 27, 917-923.

[99] Vanhaecke, L., Gowik, P., Le Bizec, B., Van Ginkel, L., Bichon, E., Blokland, M. & De Brabander, H. F. (2011). European analytical criteria: past, present, and future. *J. AOAC Int.*, 94, 360-372.

INDEX

A

acetic acid, 51, 52, 53, 54, 55, 56, 57, 58, 59, 60, 66, 71

acetonitrile, 2, 45, 47, 48, 49, 50, 51, 52, 54, 55, 56, 57, 58, 59, 60, 66, 67, 71, 72, 75

acid, 11, 12, 32, 33, 48, 49, 50, 51, 52, 53, 55, 56, 57, 58, 59, 60, 61, 66, 67, 72, 88, 93

acylation, 12

adverse effects, 15, 88

AFB1, 62

aflatoxin, 65, 81, 88

Africa, 13

age, 14, 17

agonist, 31

alimentation, 44

alkaline hydrolysis, 38

alkaloids, 84

alters, viii, 28, 35

amino, 2, 12

ammonia, 52

ammonium, 47, 52, 53, 54, 55, 56, 57, 58, 59, 60, 61, 66

ammonium salts, 66

anencephaly, viii, 27, 29

animal consumption, vii, 2

animal disease(s), viii, 14, 42, 43

antibody, 46

antigen, 46

apathy, 13

apoptosis, 12, 36, 37, 68

aqueous solutions, 2

Argentina, 4, 5, 6, 16, 19

Asia, 19

asparagus, 46, 66

assessment, 10, 15, 17, 18, 20, 88

ataxia, 13

atmospheric pressure, 74, 90

atrophy, 11

authorities, 75

B

B1 (AFB1), 65

bacterium, 10

base, 29, 31, 33, 34, 37, 67, 82

BD, 28

beer, 4, 8, 16, 21, 62, 91, 96

Belgium, 4, 5, 18, 88

bias, 80

bioavailability, 11, 15

biological samples, 84

biomarkers, 29, 48, 65, 67, 75, 89, 93

biosynthesis, 12, 29, 39

births, 34, 35

black tea, 43

blood, 12, 35

BMI, 35

body weight, 33, 34

brain, 31
Brazil, 1, 4, 5, 6, 13, 17, 20, 25
breast milk, 16
breeding, 10

C

calcium, 30, 36
calibration, 66, 71, 81, 82
Cameroon, 5, 18, 94
cancer, viii, 12, 13, 42, 87
candidates, 83
carcinogen, viii, 29, 42, 43
carcinogenesis, 94
carcinogenic, viii, 13, 27, 28, 37, 42, 43
carcinogenicity, 13, 25, 36
cascades, 32
cell death, 30
cell line(s), 13
cell membranes, 30
cellular regulation, 12
central nervous system, 14
ceramide, 12, 29, 30, 31, 36, 68
cereals, vii, 1, 2, 6, 10, 15, 18, 19, 20, 21, 22, 62, 68, 73, 93, 94, 97
challenges, 84
cheese, 8
chemical, 3, 29, 30, 83, 84, 87, 90
chemical structures, 3
Chile, 5
China, 4, 5, 6, 7, 13, 15, 25, 34, 38
Chinese medicine, 95
chitosan, 22
cholesterol, 32
chromatograms, 64, 67, 69
chromatography, ix, 42, 44, 90, 96
chyme, 24
cleaning, 11
cleanup, 84, 92
climate(s), 14, 44
clinical syndrome, 13
closure, viii, 27, 29, 37
cocoa, 58, 62
coffee, 58
Colombia, 41

colonization, 10, 22
color, iv
commercial, 16, 20, 66
community, 86
competitors, 10
complexity, 12
compliance, 83, 85
complications, 8
composition, 23, 45, 82
compounds, ix, 2, 10, 42, 43, 45, 63, 66, 67, 73, 79, 82, 83, 84, 85, 92
conception, 29, 35
congenital malformations, viii, 27
Congress, iv
consumption, vii, viii, 1, 8, 13, 28, 35, 44, 45
consumption rates, 44
contaminated food, 3
contamination, vii, ix, 1, 3, 8, 9, 10, 11, 13, 14, 16, 18, 19, 21, 22, 25, 28, 42, 44, 64, 66, 80, 84
cooking, 11
copyright, 63, 64, 65, 67, 69, 70, 71, 74, 78, 81
correlation(s), 19, 39
cost, 69, 84
Croatia, 6, 27
crop(s), 8, 9, 10, 23, 43
crown, 33
cultivation, 9
culture, 24, 30, 37, 47, 63, 79, 83, 92, 96, 97
culture media, 47, 63, 92
cycles, 45
cytochrome, 29
Czech Republic, 4, 5, 7, 16, 91

D

data analysis, 84
deaths, 33
decomposition, 9
deconvolution, 97
defects, vii, viii, 1, 13, 25, 27, 28, 29, 31, 36, 37, 38, 39
deficiency, 32

derivatives, 11, 80, 83
detection, ix, 15, 19, 42, 44, 63, 66, 75, 83, 84, 88, 89, 94
detection system, ix, 42
developed countries, 44
deviation, 80
diet, 11, 13, 15, 17, 18, 21
dietary intake, 17, 18, 44, 65
diffusion, 72
digestion, 15
diseases, vii, 2, 8, 37
disorder, 12
dispersion, 21, 73
dissociation, 83
distribution, 11, 32, 72, 97
diversity, 84
DNA, 12, 13, 24
down-regulation, 31
drought, 9, 10
drying, 8, 11, 43, 58

E

economic losses, vii, 1, 8
ELISA, 25
embryogenesis, viii, 27
embryos, viii, 32, 33, 34, 36, 39, 42, 43
embryotoxicty, viii, 28
energy, 83
environment(s), 2, 10, 21
environmental factors, 35
enzyme(s), 12, 13, 29, 30
epidemic, 38
epidemiology, 36
equine leukoencephalomalacia, vii, viii, 1, 13, 27, 28, 39, 42, 43
ESI, 47, 48, 49, 50, 51, 52, 53, 54, 55, 56, 57, 58, 59, 60, 61, 63, 67, 68, 69, 70, 71, 72, 73, 74, 75, 78, 79, 80, 92, 97
esophageal cancer, vii, viii, 1, 13, 25, 29, 41, 43
ethnicity, 29
ethyl acetate, 53, 55
EU, 79
Europe, 14

European Commission, 88
European Union, ix, 42, 44, 79, 86
evaporation, 46, 54
evidence, viii, 12, 27, 29, 35, 42, 43, 83, 86
examinations, 14
experimental condition, 29
exposure, vii, viii, 1, 10, 12, 13, 15, 16, 17, 18, 20, 24, 25, 27, 29, 31, 32, 33, 34, 35, 36, 37, 64, 65, 88
extraction, ix, 19, 20, 21, 42, 45, 46, 48, 50, 52, 53, 55, 62, 64, 65, 66, 70, 71, 72, 75, 79, 87, 89, 90, 93
extracts, 58, 71, 84, 94
extrusion, 63

F

F. moniliforme, viii, 27, 28
F. verticillioides, viii, 9, 10, 27, 28
farm animals, viii, 18, 27
FDA, 11, 24, 88
feces, 29
feedstuffs, 8, 19, 43
flight, 82, 96, 97
flora, 29
flour, 5, 6, 11, 19, 20, 69
fluid, 45
fluid extract, 45
fluorescence, ix, 42, 44, 89, 90, 94
folate, viii, 13, 28, 32, 33, 35, 37, 38
folic acid, 28, 32, 33, 35, 38
food, vii, viii, ix, 2, 4, 5, 6, 8, 10, 12, 13, 14, 15, 16, 18, 20, 21, 24, 27, 28, 29, 34, 35, 38, 42, 43, 44, 45, 48, 59, 63, 64, 69, 73, 75, 81, 82, 85, 86, 87, 88, 89, 90, 94, 95
Food and Drug Administration, 11, 24, 88
food products, 10, 20, 88, 89
food safety, 10, 69, 87
forebrain, 33
formation, viii, 10, 38, 42, 43, 63, 79
formula, 2, 30, 58, 62
fragments, 80, 84
France, 5, 18
fruits, 4, 17, 21
fungal infection, 43

fungi, 9, 10, 19, 22, 87, 88
fungus, 11
Fusarium, vii, viii, 1, 2, 3, 9, 10, 11, 14, 15, 16, 17, 18, 19, 20, 22, 23, 24, 25, 27, 28, 41, 43, 75, 79, 83, 86, 87, 88, 90, 92, 94, 96, 97
Fusarium proliferatum, vii, viii, 1, 2, 9, 22, 41, 90
Fusarium verticillioides, vii, viii, 1, 2, 9, 22, 23, 24, 41, 43, 79, 83, 94, 96
fusion, 11, 29

G

gastrointestinal tract, 29
gene expression, 22
genetics, 36
genus, vii, 2
Germany, 4, 5, 18, 66, 90
gestation, 32, 33, 34
ginseng, 19
glucose, 63
growth, 9, 10, 11, 13, 14, 22, 23, 31, 32, 33
Guatemala, 34, 37
guidance, 88
guidelines, 44, 83

H

hair, 16
harvesting, 8, 10, 23
health, vii, 1, 8, 15, 22, 39, 44, 87, 88
health effects, vii, 1
health problems, vii, 1
hepatoma, 13
hepatotoxic, viii, 27, 28
hepatotoxicity, 11, 33
herbal medicine, 24, 37
hexane, 50, 53, 58, 71
histone, 13
homeostasis, 36
homocysteine, 38
homogeneity, 80
horses, viii, 27, 28, 42, 43

host, 9
human, vii, viii, 1, 2, 8, 12, 13, 14, 15, 16, 27, 29, 32, 37, 39, 44, 55, 65, 75, 88
human health, 8, 14, 29, 44
hybrid, 20, 75, 79, 82, 96
hydrocephalus, 32, 33
hydrogen, 57, 59
hydrolysis, 11, 36, 83
hydroxyl, 2, 43
hydroxyl groups, 2, 43
hypersensitivity, 13

I

identification, 16, 80, 82, 83, 85
identity, 14, 79, 82, 83
image, 21
image analysis, 21
in utero, 36
in vivo, 24, 25, 37
incidence, viii, 9, 13, 25, 32, 33, 34, 41, 43
India, 7
industrial processing, 11
infants, 17, 34
infection, 10, 22
ingestion, 11, 28
inhibition, 12, 23, 29, 31, 32, 33, 68
inoculum, 9
integrity, 11, 13, 45, 87
intestinal tract, 24
intestine, 11
ionization, ix, 19, 42, 74, 78, 81, 82, 86, 90, 92, 96, 97
ions, 75, 80, 82, 83, 84, 85
Iran, 7, 13, 25
Ireland, 7
isolation, 87, 96
isomers, 60, 73, 83, 96, 97
isotope, 91, 95
issues, 10
Italy, 4, 5, 6, 7, 17, 21

J

jejunum, 13, 15
Jordan, 7, 21

K

Kenya, 23
kidney(s), vii, 1, 11, 12, 28, 33, 38
kidney tumors, vii, 1, 28

L

laws, vii, 1, 11
LC-MS, ix, 18, 21, 24, 42, 44, 46, 47, 52, 62, 64, 66, 67, 68, 72, 73, 74, 75, 79, 82, 86, 89, 90, 91, 94
LC-MS/MS, 18, 21, 24, 46, 62, 63, 64, 66, 67, 68, 72, 73, 75, 79, 82, 86, 89, 90, 91, 94
lead, 12, 13, 85
legislation, 62
lesions, 11, 13, 68
lethargy, 33
life sciences, 85
Limpopo, 34
liquid chromatography, vii, ix, 17, 18, 19, 20, 21, 42, 44, 62, 65, 69, 82, 86, 88, 89, 90, 91, 92, 93, 94, 95, 96, 97
Liquid Chromatography-Tandem Mass Spectrometry, 92
liver, vii, viii, 1, 11, 12, 13, 24, 25, 27, 28, 29, 32, 33, 46, 50, 64, 90
liver cancer, viii, 25, 27, 29
locomotor, 13
Luo, 19

M

maize, vii, viii, 1, 2, 3, 5, 7, 8, 9, 10, 11, 12, 13, 14, 15, 16, 17, 18, 19, 20, 21, 22, 23, 27, 28, 29, 34, 35, 37, 41, 43, 44, 46, 48, 68, 69, 73, 75, 80, 83, 86, 88, 89, 90, 92, 95, 97
Maize, vii, 1, 2, 4, 5, 6, 7, 22, 48, 50, 51, 52, 53, 55, 60, 93, 94
majority, 44
Malaysia, 7
management, 9, 24
mass, vii, ix, 17, 18, 19, 20, 21, 42, 44, 47, 48, 49, 50, 51, 52, 53, 54, 55, 56, 57, 58, 59, 60, 61, 63, 64, 65, 66, 68, 70, 72, 75, 79, 82, 83, 84, 85, 86, 89, 90, 91, 92, 93, 94, 96, 97
mass spectrometry, vii, ix, 17, 18, 19, 20, 21, 42, 44, 65, 82, 83, 84, 86, 89, 90, 91, 92, 93, 94, 96, 97
materials, 46, 62, 68, 73
matrix, 16, 21, 24, 45, 66, 73, 75, 80, 81, 82, 84, 85
measurement(s), 82, 83, 84, 85, 86
meat, 8, 72, 93
media, 22
Mediterranean, 7, 21
Mediterranean countries, 7
metabolism, viii, 12, 13, 24, 28, 29, 30, 31, 32, 33, 34, 35, 37, 48, 67, 75, 89
metabolites, 24, 28, 30, 31, 43, 46, 50, 52, 53, 57, 85, 90
metabolized, 8, 29
methanol, 2, 45, 46, 47, 48, 49, 50, 51, 53, 54, 56, 58, 59, 60, 61, 66
methodology, 17, 18, 91
methylation, 13
Mexico, 34, 37, 41
mice, viii, 13, 25, 28, 31, 32, 33, 38, 39, 42, 43
midbrain, 33
mineral water, 4
MIP, 46, 51
mitosis, 12
moisture, 11
moisture content, 11
molds, vii, viii, 27, 28
molecular structure, ix, 42
molecular weight, 2, 84
molecules, ix, 42, 68, 73, 83, 84

Morocco, 5, 17
morphogenesis, 32
myelomeningocele, 29, 37

N

NaCl, 53, 56, 57, 58, 59
naphthalene, 24, 37, 89
natural compound, 15
necrosis, 13
neonates, 36, 38
nephrotoxic, viii, 27, 28
Netherlands, 5
neural tube defects, vii, viii, 1, 13, 25, 27, 28, 29, 36, 37, 38, 39
neural tube development, viii, 37, 42, 43
neurogenesis, 31
neurons, 13
New Zealand, 5, 33, 37
Nigeria, 6, 19
NIR, 19
nitrogen, 46
non-polar, 2, 73
NTDs, viii, 27, 29, 31, 32, 33, 34, 35
Nuevo León, 41
null, 31

O

Oceania, 6, 19
ODS, 48, 55, 60, 83
oil, 5, 18, 23, 58
optimization, 17, 94
organic food, 88
oxidative stress, 30

P

Pakistan, 22
paresis, 13
pathways, viii, 28, 79
permission, 63, 64, 65, 67, 69, 70, 71, 74, 78, 81
permit, 86

pesticide, 19
pests, 10
pH, 47, 52, 57, 66, 84
pharmaceutical, 85
phenolic compounds, 22
phosphate(s), 12, 31, 32, 37, 60, 67
phosphorylation, 31
physicochemical properties, 84
pigs, viii, 27, 28, 42, 43
placenta, 32, 33
plant growth, 9
plants, 22, 43, 68, 88
playing, 45
plexus, 13, 15
Poland, 6, 19
polar, 2, 45, 73
polymer(s), ix, 42, 46, 87, 89
population, viii, 15, 20, 29, 41, 43, 44, 88
population group, 88
Portugal, 4
precipitation, 9
pregnancy, 29, 33
preparation, ix, 17, 28, 42, 45, 62
prevalence rate, 34
prevention, 9, 32
processing stages, 10
project, 87
proliferation, 68
propane, 2
public health, 43
pulmonary edema, vii, viii, 1, 13, 27, 28, 42, 43
purification, 45, 55, 64, 93

Q

quality control, 68
quantification, 7, 19, 66, 71, 80, 84, 85, 88, 89, 96, 97
questionnaire, 35

R

reagents, 89

real time, 85
receptors, 31, 32
recovery, 9, 46
reducing sugars, 63
regions of the world, viii, 27, 29, 34
regression, 15
regression model, 15
regulations, 24
rejection, 11
reproduction, 34
requirements, 45, 85
residues, 9, 19
resistance, 22, 72
resolution, ix, 20, 42, 45, 69, 80, 82, 83, 85, 86, 96, 97
respiration, 36
response, 75
retardation, 33
risk(s), 21, 23, 24, 25, 32, 35, 36, 37, 44, 87
risk assessment, 21, 24, 36
risk management, 23, 87
rodents, 13, 28, 38
root, 9, 23
root rot, 23
rural areas, 34

S

safety, 44, 69
scope, ix, 42, 83
seed, 10, 22
seedlings, 48, 68, 75, 89, 95
selectivity, 46, 72, 73, 79
sensitivity, 72, 75, 79, 82
Serbia, 4, 6, 16
serum, 35, 38
shape, 73
showing, 46, 64, 73
side chain, 80
signal transduction, viii, 28, 30
signaling pathway, 31
signals, 83, 84
silica, 72, 73, 74, 95
silk, 22
skin, 32

small intestine, 13
socioeconomic status, 35
sodium, 57, 59, 61
software, 85
solid phase, 93
solubility, 84
solution, 10, 28, 65, 67, 71, 72, 75
solvents, 2, 45, 66
South Africa, viii, 5, 7, 25, 34, 36, 38, 41, 43
soy bean, 43
soybeans, 8
Spain, 4, 5, 6, 7, 18, 22, 41, 87, 88
species, vii, 2, 9, 10, 11, 13, 14, 16, 21, 43
specifications, 11
spectroscopy, 19
sphingolipid metabolism, viii, 12, 13, 28, 29, 30, 31, 32, 33, 35, 37, 48, 67, 75, 89
spina bifida, viii, 27, 29, 37
Sri Lanka, 5, 18
stability, 80, 84
standard deviation, 45, 66
state, vii, ix, 42, 86
storage, 8, 9, 10, 11, 23, 43
stress, 9, 10
structure, 9, 11, 22, 29, 30, 43, 63, 84
styrene, 24, 37
substrate(s), 2, 8
sucrose, 63
supplementation, 32, 35, 38
suppression, 81, 84
survival, 9, 30, 31
Switzerland, 87
symptoms, 13
syndrome, 28
synthesis, 12, 31
Syria, 4, 17

T

Tanzania, 4, 5, 16, 17, 18
target, 45, 81, 82, 84, 85, 87
techniques, vii, ix, 9, 42, 45, 74, 75, 82, 86
technology, 75, 95
temperature, 28

teratogenicity, viii, 28, 35
thermal stability, 8
threshold level, 35
tissue, 30, 32
toxic effect, 14, 28
toxicity, viii, 11, 12, 15, 22, 24, 27, 30, 32,
    33, 34, 36, 38, 42, 43, 86
toxicology, 15, 96
toxin, 8, 11, 21, 65
transport, 12, 32, 37
treatment, ix, 10, 22, 28, 30, 32, 33, 42, 45,
    47, 48, 49, 50, 51, 52, 53, 54, 55, 56, 57,
    58, 59, 60, 61, 62, 86, 96
tricarboxylic acid, 2, 80
trifluoroacetic acid, 47, 66
tumorigenesis, 13
tumor(s), 13, 28, 38
Turkey, 7
turnover, 31

U

UK, 95
underlying mechanisms, 12
United Kingdom, 5, 14
United Nations, 12, 24
United States, 5, 14

urban, 34, 36
urine, 12, 46, 55, 65, 75, 91, 93
USA, 7, 95
UV, ix, 42
UV light, ix, 42

V

validation, 17, 91, 94
varieties, 16

W

Washington, 14
water, 2, 6, 9, 11, 16, 22, 45, 47, 48, 49, 50,
    51, 52, 53, 54, 55, 56, 57, 58, 59, 60, 61,
    66, 67, 71, 72, 75, 80
white matter, 13
WHO, 37, 87
World Health Organization, 87
worldwide, viii, 27, 43

Y

yolk, 32